Tracing The Trajectory
Career Experiences Around The Globe

Class of 1986
PSG College of Technology

TESTIMONIALS

It is quite rare that those who get a degree return something that is intellectual to the institution at which they have studied. It is even more rare that a class of students, which has achieved distinction in a wide range of fields, would come together to share their experiences for the benefit of those that follow.

This is exactly what the class of 1986 has done. Their combined knowledge and experience will help not only the students of PSG College of Technology, where they all studied engineering thirty years ago, but students elsewhere as well. As an alumnus myself, I am delighted to heartily congratulate them on their unique contribution.

Dr. Raman Kolluri
Fulbright Scholar
Head of Department
Physics and Materials Science
PSG College of Technology (1970-85)

I want to congratulate the graduates of PSG College of Technology, class of 1986 for their innovative, informative, insightful and inspirational gift to the current and future students of their alma mater. The impact of this gift will likely last for a long time to come. As I read this book, I longed to be 18 again, with a renewed appreciation for the cocktail of curiosity, perseverance, passion, risk seeking and serendipity!

Dr. Narendra Agrawal
Benjamin and Mae Swig Professor
Department of Operations Management and Information Systems
Leavey School of Business
Santa Clara University

TESTIMONIALS

The 1986 class at PSG College of Technology has launched this book as a worthy effort at contributing back to its alma mater. Firstly, the panoramic view based on their experiences years ago on campus, and their current pursuits, is valuable for students who are studying there. Secondly, it shines the light on the thoughts of young talent from India that is now spread out either within the country, or as Indian diaspora, across the globe.

I enjoyed and took away nuggets of knowledge that escaped my attention as a chemical engineering major at the Indian Institute of Technology in Mumbai during the latter half of the seventies. My compliments to the contributors and wish this effort every success.

Dr. Milind Shrikhande
Clinical Professor, Finance
J. Mack Robinson School of Business
Georgia State University

We are used to reading about individual accounts of travels such as Cook or Magellan. This book is fascinatingly different. A set of students, all starting at the same place and time, branch out intellectually and geographically. The result is a compilation of a wide range of experiences, thinking, and opinions that are global, contemporary and informative.

Dr. N. R. Prabhala
Head of Research
Centre for Advanced Financial Research and Learning (CAFRAL)
Supported by the Reserve Bank of India
and
Professor of Finance
University of Maryland

TESTIMONIALS

In the world of packaged educational programs, MOOCs and distance learning degrees, slogans that attempt to run universities like businesses, and where students are increasingly viewed as customers to be pleased, emerges a spontaneous expression of the love for the essence of education. The people expressing their views in this volume are united by a core bond formed years ago, when they acquired a Bachelor's degree. They left with the ability to think for themselves, to ignore geographical and disciplinary constraints that would otherwise stunt growth, and with a sense of service and a desire to make an impact in whatever they chose to do.

Having moved on to far flung places like Tennessee, the still rumbling island of New Zealand, and the frozen tundra of the northern reaches of the earth, they still trace their success to their formative years. One young student saw a metallurgical craftsman exercising the art of engineering with pride to achieve the perfect amalgam of properties validated by scientific analysis in the lab, another was inspired simply by a casual conversation with a classmate and yet another by a sympathetic ear of a teacher at a tough time. We also see here the actual embodiment of "women in engineering" well before it became a trend. This is clear evidence that the experience of a cohort going through an engineering program at a caring institution can hardly be replaced by distance learning and other "modern" concepts.

Reading these stories leaves one with the comfort of knowing that the future of engineering and its impact on society is in good hands. One can also see glimpses of India embracing once again the value of its ancient *guru-shishya* traditions, to give to the world real models of education and life-long learning. Engineering education administered in this manner is, above all, education that brings about Human Transformation.

I wish these veteran graduates of PSG College all the best as they continue to leave their mark on the world. And I am confident that this outpouring of reflections from this group of students will serve as the compass for many generations of students to come.

Dr. Srinivas Garimella
Hightower Chair in Engineering
Professor and Director
Sustainable Thermal Systems Laboratory
Georgia Institute of Technology

TESTIMONIALS

These essays from a diversity of professional and geographical locations will offer much valuable insight and inspiration to the current students of PSG College of Technology. Hindsight is 20/20, and what better way than for today's students to benefit from these alumni perspectives, as they chart out their brave new paths in the world. The reflections in this volume open up a fascinating range of experiences and advice for today's generation of students as they prepare to be tomorrow's leaders.

Prof. Kavita Daiya
Visiting NEH Chair in the Humanities, 2015-2016
Albright College

Associate Professor of English
George Washington University

I am delighted to provide a testimonial for these compilations of the 1982-86 batch of PSG Tech Alumni. I have been in touch with them, since their college days and they are a vibrant group of professionals with excellent track records.

The presentations in this book are vivid, experiential information, which facilitate the readers to connect and relate to the authors authentically. One can clearly see in the writings, how the aspirations of an individual changes, with the time and that there is a shift in their perceptions about the world. The writers have put down in a story-like manner, their understanding of life, application of the facts learned and their scripts distinctly capture their professional and emotional growth over a period of time. The discussions are full-of-life, with striking examples of their life expedition, which provide an aspiring reader valuable tips for his/her career.

Their contributions are unique and commendable.

Dr.P.V. Mohanram
Principal
PSG Institute of Technology and Applied Research
Coimbatore, India

DEDICATION

Gurus

Dr. K. Venkataraman (Principal)

Prof. Gandhimani (Mechanical)

Prof. G. Gurusamy (Electrical)

Prof. K. Jayachandran (Textile)

Prof. Jayaraman (Electronics)

Prof. Muthusamy (Mechanical)

Prof. Narayanaswamy (Electronics)

Dr. S. Palanichamy (Electrical)

Prof. Ramakrishnan (Civil)

Prof. Seetharaman (Civil)

Prof. Sundaram (Mechanical)

Friends

Mr. Dharmaraj (Electrical)

Ms. Geetha (Civil)

Mr. Karunakaran (Mechanical)

Mr. Malmarugan (Production)

Mr. Murugesan (Textiles)

Mr. Ravi (Mechanical)

Mr. Ravichandran (Mechanical)

Mr. Senthilvel (Metallurgy)

Mr. Shanmughasundaram (Textiles)

Mr. Sridhar (Production)

Mr. Venkatesh (Mechanical)

Mr. Vijayaraju (Civil)

CONTENTS

FOREWORD

When I was approached to write a foreword to this compilation of the life journeys of the PSG Class of 1986, I felt happy for a couple of reasons. I have been fortunate to closely interact with many in this class during their student days. I have also been in touch with many on and off after graduation. Graduates of this class have been progressing well professionally as well as getting recognized by the communities where they live, through their contributions.

In the past, the employment scenario for engineers was characterized by lifelong loyalty to their organization, with mobility within the confinements of geographical boundaries of the organization spread. In addition, there was a lifelong adherence to a basic discipline, though shifting from a technical stream to a managerial stream in the course of time was the norm.

However, going through the stories and the journeys in this volume, I find that information technology has brought about a paradigm shift. These graduates have successfully refocused their core competencies depending on professional opportunities they identified, which change continuously and are often not at all related to basic specialization at the undergraduate level.

Mobility across borders for employment has now become the norm rather than the exception, as this offers opportunities for better employment along with professional challenges and a better quality of life. Technological developments have enabled large scale, across the border movements of graduates for employment and entrepreneurship. In many cases the basic undergraduate qualification becomes only a stepping-stone to take advantage of whatever opportunities come in the way.

As an educator, it occurs to me that this is also a pointer for designing an education system that should make graduates capable of adapting themselves to fast changing technology and business scenarios.

We rejoice in the success and achievements of the graduates of this class. It gives us confidence that the current and future generation of our alumni also will follow the path shown by the trailblazers such as these. My best wishes for the continuation of these journeys to excellence. And best wishes to those that are beginning it in this changing world!

With warm regards to all of the 1982-86 alumni from the PSG College of Technology.

Dr. Radhakrishnan
Director
PSG Institute of Advanced Studies
Coimbatore, India

INTRODUCTION

Origin of an idea

In December 2005, exactly ten years prior to the publication of this book, a group of three classmates of ours started a virtual discussion group out of the blue. We were not sure what they expected when they initiated it. The response was overwhelming and immediate.

It was as though twenty years of latent demand that had been bottled up was suddenly uncorked, and the spontaneous joy of interaction was contagious. Hundreds upon hundreds of messages and greetings poured in from all corners of India and the world.

An observer of this gush of enthusiasm might be forgiven for thinking that this would soon fade, perhaps thinking about such experiences with other discussion groups. However, for reasons yet to be fully explained, this discussion group went from strength to strength and today in addition to its original format, has branched into the various multi-media platforms that we are all familiar with today - be they podcasts, blogs, chats, teleconference calls or in-person meetings.

Here is one telling example. In 2014, two classmates hit upon an idea to propose one topic a week. Every week a new classmate would propose a topic. And the rest of the class would jump in with their points of view. These topics included work, family, current affairs, arts & entertainment and travel. The conversation was so voluminous that we turned it into an informal book that we shared with the entire class. This effort in 2014 was indeed the origin of the idea for this book.

Realizing what we had just done as a class in 2014 for our own benefit, we wondered could we not take the same energy and direct it towards creating a book for the graduating class of 2016 and beyond? The response again was immediate and enthusiastic. Thus 2015 was spent meeting, planning, writing, editing and publishing this book.

Why?

Any endeavor, and the explanation of any endeavor must always start with the question - *Why?* What is the purpose of the endeavor? If the question *why?* is answered, then the questions *what? who?* and *how?* will automatically follow.

The rationale for the book is this: When we were 22 years of age and just graduating out of college, we were not at all aware of how much the world was changing and what we needed to do to navigate this world. And indeed, when we look back thirty years, we are thankful that we have been able to navigate reasonably well thus far both in our personal and professional lives. We have had great teachers, mentors and managers along the way.

What would it be like, we wondered, if we were 22 years old again, and a group ahead of us distilled their experience and shared it with us? Not as a recipe or a formula or even a guideline, but rather as a set of observations, made in an open minded and flexible way that could shine a light, however faint on the path ahead.

We are inspired by a couplet in the *Upanishads* that says, "In the dark of the night when lightning strikes, you see the world as it is for a brief second. You hurriedly record what you see." It occurs to us that you need to be awake when the lightning strikes and be ready to 'see' when it occurs and then have the talent to record it!

We believe we have been awake to our collective experiences. And we have 'hurriedly' made notes of it. We wish to share these notes with you.

Who?

Ostensibly we have written this for the graduating class of 2016 and beyond at the PSG College of Technology, our alma mater. This is because we have walked in your shoes on that very campus. Thus, one might imagine that we relate to you most immediately.

However, our hope is that this book is for students of the graduation class of 2016 and beyond elsewhere as well, since these experiences are in a multitude of careers and span geographies across the globe.

What?

There are nearly 20 chapters in this book. Each chapter is on a different discipline or subject. There are chapters on Materials Science, Oil & Gas Exploration, Earthquake Engineering, Information Technology, Financial Services, Entrepreneurship and a variety of other topics including reflections on how to navigate a career or professional life in a new setting. Each chapter follows a general guideline:

- Why did the author choose his or her field?
- What was the career trajectory?
- What were the career choices?
- What is their view on the future of their field?

Finally, each chapter includes any advice and overall observations that might be of interest. We have kept these chapters brief, personal and hopefully engaging.

Between chapters, we also have brief observations on leadership and career management made by a select group of classmates as well in sections titled, *"Question & Answer."* We have asked which leader has inspired our classmates in their careers and in their lives overall.

A disclaimer – the contents of each of these chapters are personal views of the authors and do not represent the positions of their respective organizations.

How?

This book is not constructed so that it is necessary to read it from the first page to the last. However, you are free to do so!

This book is constructed in such a way that any chapter, or any observation should be able to stand on its own.

- If a particular topic is of interest (say, Earthquake Engineering), you could jump directly to it.
- If a particular geography is of interest (say, New Zealand), you could go there.
- If you would like to read a chapter by a female author, rather than a male author, you could do it.
- If you want to read a general chapter such as the one on entrepreneurship, rather than a specific topic in engineering, you could choose that.

In other words, you could choose to navigate to any part of the book that most immediately appeals to you.

Contacting Us

We encourage you to contact us with your comments and questions at the following email address: TracingTheTrajectory@gmail.com.

A concluding thought

We have labored to create this book in the hope that you are inspired by our observations of this fast changing world. And we hope that you find navigating your future just a little easier.

The Return On Investment (R.O.I) of our effort, we hope, is that a few inspired minds will make a positive contribution back to our alma mater.

Our warmest regard and best wishes for your successful future.

Class of 1986
PSG College of Technology
On the occasion of our 30th year meet
Dubai, UAE

Surya Kolluri
Mechanical Engineering
General Editor
Boston, USA

December 31, 2015

I. THE ENGINEERING LIFE

MATERIALS SCIENCE AND ENGINEERING

Pursuit of Material World(s)

Tracing a path from India to the United Kingdom to the United States of America, the author shares his insights about new materials such as graphene and techniques such as 3D-printing.

Introduction

Thirty years after graduating from the PSG College of Technology and looking back, I am humbled by our teachers' passion and am realizing the real meaning of the kriti, *"Endharo mahavubhavulu, andhariki vandhanumu,"* composed by Saint Tyagaraja. In the same spirit, I salute the great teachers and scientists, as well as, my friends who have inspired me to pursue the world of materials science.

This chapter provides an overview of progress and future direction of shaping the materials in the last thirty years from my viewpoint. I hope these anecdotes and thin slices of experiences strengthen the creativity and grit of future students.

The direction of pursuit is cast

On a hot summer afternoon thirty years back, the train named Kovai express was making its usual last lap of the journey, its wheels hugging the steel rails near the *Peelamedu - Avarampalyam* curve. I looked out through my window with trepidation of my future at the PSG College of Technology. On either side of the tracks, I saw lot of small factories with tubular furnaces among grey sand piles.

I turned around asked my brother-in-law, who was traveling with me, what these were. He looked at me with pride, and said that those are *cupola furnaces* in foundries, the economic engine of Coimbatore. With beaming eyes, he further explained that they make parts that go into all other machines, which are used by other business sectors, including textiles, another big sector in the city of Coimbatore. Somehow, as the air with burnt iron fragrance enveloped me, I knew that my *life was cast* in the materials world.

As the time at college moved on and we made friends, as we learned about engineering tools, to design parts, melt metals, make molds for castings, blacksmithing, machining, and welding, our tapestry of life was unfolding right in front of us.

Like a cooking recipe

While at college, a technician in-charge of melting furnace at PSG College of Technology foundry greatly influenced me. All day long, he would take a scoop of molten metal cast it into a disc shape. He would put that disc in a pneumatic tube. Magically, it would be transported to another part of the building. Then, he would wait for a phone call for a report that helped him to make the decision to pour or not to pour the molten metal into casting.

I approached this man and asked him, what he was doing. He looked at me said, *"hey kid, this is like cooking, if your ingredients are not in proper ratio, your food will be bad."*

If the molten material was not in the right composition, all the castings would be junk and the foundry will lose lot of money. I could see his passion, pride and commitment to his job.

Reluctantly, not hiding my lack of knowledge, I asked what was at the end of that tube. He said there is a lab, which checks the recipe. He followed with an authoritative voice in Tamil: *"Book knowledge can only take you half the way; to scale the whole mountain, you need to have the practical knowledge of using ropes and ladders."*

Question and Answer

With my friends Sundar and Subbu, I made the trek to the laboratory and found the plasma spark analysis instrument. That machine took those discs and evaporated the constituents into a vapor and measured the relative concentration of silicon, chromium, molybdenum, nickel and iron.

Again with reluctance, I asked why these recipes were important for the casting. The lab manager looked at us with a weird look and said, *'you are wasting my time, you should have learnt it from NKS.'*

NKS, aka., Professor N. K. Srinvasan taught us *physical metallurgy*. In one of the lectures, while he was talking about dislocations and its impact on properties, he articulated, with a glitter in his eyes, his research experience pertaining to semiconductor materials at *Columbia University*, in New York in the United States of America. I was convinced, at that moment, he was not in that classroom; he had time-transported himself to his past to a lab in

4

Columbia.

This was the moment I realized there is a fun in asking questions, *"why, what, when and how - related to materials world,"* and can also build one's career answering the questions.

Pursuit of stronger materials
(aka the Complex life of Fe and C atoms)

I was a fan of comic books when I was growing up. I used to wait at the street corner shop for my favorite *"Irumbukai Mayavi"* published by Muthu Comics. This is based on British superhero, who has an artificial hand. On plugging his hand into A/C circuit, he goes through extreme pain and gets charged up, and becomes invisible. At the same time, I used to hear about another superhero *"Iron Man,"* published by Marvel Comics too.

I always wondered when I would get this magical iron claw that can drive off the demons that used to haunt me after my father passed away. The mystical material iron really fascinated me.

In my 2nd year, at college, most of us were really taken aback by the statement by Professor Ramakrishnan: *"you can take pure iron and bend it with your hand. But on addition of small quantities of pixie-dust, you can make it stronger. Do you know what that pixie-dust is?"*

He went on to explain us the role of carbon and the background to steel and how it plays its complicated dance of being close to iron atom at high temperature and heating the iron atom at low temperature, due to the duality of iron atoms in arranging themselves in space, explained by crystal structures.

In preparation for being metallurgical engineers, we studied the ELBS books on Physical metallurgy by Avner and Mechanical metallurgy by Dieter. We memorized the *cooking recipes* to make wide range of steels. We would get confused by different names given to arrangement of crystals and phases after people like Austen, Marten and Bain.

A question always remained in our minds, why are humans blessed with this magical recipe of Iron and Carbon which allows us build huge skyscrapers like Eiffel Tower, Empire State Building, cars like Ambassador and other engineering marvels?

Onwards to Cambridge University

The answer was in lots of textbooks and papers. But it is nothing like hearing it from a passionate researcher of steel. Dr. Harry Bhadeshia had moved to the United Kingdom as an immigrant from Kenya. On arrival at Cambridge, while I was being frustrated with my inability to understand complex equations and theories of steels thrown at me, Harry asked me to join him for a nice lunch at Darwin College.

As we walked and talked about everything in life, I asked him how he became smarter with all these equations. He paused for a moment and said, there is no short cut; you needed to work hard, understand each sentence you read and most importantly you should visualize in your mind the concepts.

Then, I asked him how he came up with the theory of bainite transformation. He told me that with his usual stock phrase, *"Look, Suresh! We all stand on the shoulders of master."* He went on further and showed his collaboration with Prof. Jack Christian from Oxford, who had proposed this possibility before him. He just took it to the completion with emerging computational tools and characterization.

I can see that Harry will never stop until he makes the best steels known to us by manipulating the arrangement of carbon atoms in iron lattice so that they are cheaper, formable, high-temperature resistant, weldable, and wear resistant. Well, even now he is continuing this journey with so many.

Suddenly, I realized that scientists are not born; they evolve from simple humans by asking questions *Why? When? What? How?* work hard and keep trying until they reach the end of the rabbit hole and keep digging." That means a student, including the one who is walking on *Avinashi road* in Coimbatore can be a scientist, the degree you receive is just a driving license to do research!

Innovation

In 2005, while attending a steel conference at a monastery in Kyoto, Japan, a young entrepreneur, Gary Cola, approached me, *"I have made steel stronger in 30 millisecond better than anyone in the world, are you interested in working with me?"*

Usually, faculty who are engrained in pedagogy would dismiss him. However, I remember what Harry taught me, be mindful of innovation and creativity lurking around the corner, irrespective of degrees you hold. This was the most challenging and exciting part of research my students and I did, where we discovered how Gary's process works and how we can join them too.

Interestingly, we live in the world, where carbon alone wants to rule the world without the iron, i.e., discovery of *graphene*, which is claimed to be 100 times stronger than steel. Now, I wonder whether this divorce of iron and carbon will lead to better world in the long run?

I hypothesize that we will continue to make stronger and better and cheaper steels for macro world in which iron and carbon will live happily together in *FCC* crystal structure and hating each other in *BCC* crystal structure from time to time. At the same time, a nano-world will emerge where iron and carbon atoms may go on their separate way to create exciting electromagnetic devices!

Now, you may wonder, do we have a way to create meta-materials that may bend the light and make us vanish and be stronger, like *irumbukai mayavi* and *Harry Potter*

Remember that the answer lies around us. Thus we need to ask the question and persistently search for an answer.

I suggest that we should not wait to get a driving license to conduct research; we should ask the why, what, how questions now.

Pursuit of stronger bonding between materials (and peoples)

One of the questions of life in real world we face is following: how can we address challenges that are far bigger than an individual can make it happen?

I remember as a kid watching the movie *Pallandu Vazhga*, with lead role by the legendary MGR. In this movie, as a jail warden, he takes up the challenge of rehabilitating six inmates. The warden's motivation is questioned by everyone, asking that how helping six inmates going to make a difference to the whole country. In a clever, but relevant, demonstration is done in the movie. He draws a figure of a human behind a map of India. Then he tears the map, and shuffles and asks his audience to arrange it using the map-face side of the paper. All of them have trouble in fixing them. All of them are frustrated to certain extent, at that time, he asks them to turn the map around and arrange the human figure. All of them can do it very easily. When the paper is flipped, the audiences get the whole map of

India.

This can be a metaphor to translate into the materials world. In the late 1980s, an engineer named Wayne Thomas from *The Welding Institute* was working on solid-state welding of metals and alloys. One day, he had a brilliant idea that if we can use a rotating and moving hard tool that can plasticize metals in the solid state and interpenetrate them by cross flow, we can make large-scale solid-state joined structures. He postulated that it could be used for joining high strength aluminum alloys used by aerospace industries, which are very difficult to join by fusion welding processes.

No one took his idea seriously.

However, he did not listen to the naysayers, since he was confident that his idea would work. He independently worked and developed this technology called *Friction Stir Welding*. Now this technology used by many industrial sectors, including the upcoming *space launch system*. This clearly shows one man's grit can indeed bring the material world together.

This story made me to devote my career to the challenging world of joining materials in three dimensions including the emerging technology referred as additive manufacturing, also known as *3D printing*.

Summary

The world of new materials is emerging with bridges being made between organic and inorganic materials with their unique properties. We need to design hybrid materials with form and functions by shaping them similar to nature while dealing with diversity and heterogeneity of their properties for betterment of society around us. In Mahatma Gandhi's words, we should aspire to *"be the change that you wish to see in the (material) world."*

About the author:
Suresh Babu obtained his bachelors degree in metallurgical engineering from PSG College of Technology and his master's degree in industrial welding metallurgy from Indian Institute of Technology, Chennai. He has a PhD in materials science and metallurgy from University of Cambridge, UK. He currently occupies the UT/ORNL Governor's Chair of Advanced Manufacturing at the College of Engineering at the University of Tennessee in Knoxville, TN.

Question & Answer

Which leader do you admire the most?

My favorite leader is U.S. President Abraham Lincoln. He was a great leader, thinker and speaker. It was a proud moment for me when I had a chance to visit the Lincoln Memorial recently and I spent a lot of time around and took pictures standing near the statue.

Lincoln's most luminous leadership insight was, "*In order to win a man to your cause, you must first reach his heart, the great high road to his reason.*"

Lincoln fundamentally cared about people and made every effort to demonstrate that to them. He was in essence a genuine human being who identified with the challenges people faced and the sacrifices they made.

Lincoln's prodigious influence on friends and foes alike was due to his extraordinary empathy - the ability to put himself in the place of another, to experience what they were feeling and to understand their motives and desires.

I recommend the book *Team of Rivals* by Doris Kearns Goodwin.

Kanchana
Production Engineering
Mumbai, India

Note: The *Question and Answer* section appears throughout the book, in-between chapters. These explore topics such as sources of inspiration and choices that have impacted our careers and lives.

CIVIL ENGINEERING

ENGINEERING FOR EARTHQUAKES

Tracing a path from India to New Zealand, the author shares her experiences on how she ended up in this field and how the best minds are working hard to try and protect us from devastating losses due to earthquakes.

Introduction

As a woman, brought up in India, cutting a path and career through Civil Engineering has been an adventure in choices, challenges and risks. I am a person who is comfortable taking career risks, but find often that it is not easy to fully understand the depth of the risk that I am taking.

I hope to share here the career choices I have made, the risks I have taken, how I landed with earthquake engineering as a specialization and finally how this might be of value to you.

Because my father was a Civil Engineer!

I must have scored well in my exams, since I received a letter from the board saying that I was allotted a seat at the PSG College of Technology.

I was assigned 'Civil Engineering' as my branch. I had wanted to try electronics or electrical engineering, eager to try these growing fields, which from what I had heard were supposedly more appropriate for women.

When I tried to get a different major I was told, "*your father is a civil engineer. You will be one too.*" The story sticks with me quite vividly today. However at that time I did not make much of it. I thought it was fine and applied myself to studying in that field of engineering. I should say that my father is still enjoying the memories of his achievements and accomplishments in his civil engineering career.

Civil Engineering being the second-oldest discipline and thus a mature field, there were concerns that there might not be lucrative job offers. But as I got deeper into the subject matter, I got very much-interested subjects such as mechanics, structural analysis and structural design. I did well enough that I got admitted to the graduate program at the Indian Institute of Technology (IIT) Chennai.

A passion for teaching

I joined the Regional Engineering College (REC) at Tiruchinapalli and had a teaching assignment. I found that my 'teaching career' was rewarding. I generally received very good appreciation from the students for my ability to explain things well.

However, the assignment was only for one year. I realized a PhD was essential to continue developing my teaching career. So, I pushed myself to do PhD and went back to IIT Chennai.

Awake to earthquakes

At IIT Chennai, I was exploring research areas that would be interesting/challenging about new trends in civil engineering. One professor advised me to choose problems related to 'Dynamics' rather than 'Statics".

I decided to undertake research in the field of earthquake engineering. A young and energetic Assistant Professor agreed to offer supervision. I chose a topic related to seismic assessments of buildings. The significance of this topic came clear when there was a massive earthquake in Uttarkashi in 1991. Coincidentally, that is when I registered for my PhD!

That is how I ended up getting some exposure or knowledge in earthquake engineering.

The course was new and hard for me. So many times I was considering quitting.

A friend who was in her final stage of her PhD program, provided me moral support and said: "Uma, I know it is not easy. But once the assignment is taken, stick with it and finish it, come what may!"

I did so amidst so many challenges, thanks to my husband who was so supportive and patient during the trying period. I now realize that while I was doing the PhD, I did not know the kind of openings this degree would give me in my later stages of my life.

I enjoyed thoroughly my teaching profession to deal with civil engineering subjects. Whenever I get an opportunity, I inspire young minds by showing new and innovative structures designed around the world, just to let them know how much more we can do in civil engineering.

This, I believed could help them overcome a kind of feeling that "civil engineering' was not good enough as other subjects. Also, I make a point that students should develop respect for every discipline because any discipline cannot function independently if they want to deliver something useful to the society. I served as a teacher for 5 years after my securing my PhD.

A bend in the road

My personal life then took a drastic change in direction. Given these changed personal circumstances, I decided I would take a risk and resign my job. I did not have an idea or plan on what I would do next. But I did stay abreast of recent research developments in earthquake engineering.

I had the opportunity to attend two international conferences overseas and those visits helped me to get updated in my areas of research interests (seismic assessment) as well as to get an idea how living outside India was.

One place was New Zealand. I successfully got a Post-Doctoral position at the University of Canterbury, Christchurch. My PhD topic in the field of earthquake engineering was considered to be a "well qualified degree" for New Zealand because that country has high earthquake risks.

A new country

New Zealand is a small country but faces high risks from natural hazards. The city where I live is Wellington and one of the major earthquake fault ('Wellington fault') runs across the city. And just at a distance of 250 meters from my house!

Since I moved to New Zealand in 2005, one of the major topics of discussion is around seismic safety policy for buildings and occupants. Recent mild and sever earthquake events keep the topic live and continuous attempts are made to refine the government policy to strike balance between 'seismic safety' and the economic drivers and constraints from the society.

At present I work for one of the Crown Research Institute of New Zealand where pure and applied research activities are encouraged. Applied areas of research give a sense of reward to me.

In New Zealand, work-life balance is given importance and priority is considered for family even at the work place. It is a beautiful country to raise our kids in safety.

Construction in the face of earthquakes

I have had the direct experience of mild ground shaking under my feet. However, recently a devastating series of earthquakes in Christchurch gave me a vast exposure to the types of damages to built-structures, the land and hence the functioning of community.

These impacts have raised issues to be looked at within different construction sectors, structural engineers, geotechnical engineers and social scientists to work independently and collaboratively towards community resilience.

In this context, there is a tendency to promote new types of construction that will sustain 'low damage' and the damage can be relatively repaired with minimum disruption. However, there are always challenges in adopting any type of changes and implement a system or a policy for adopting new systems.

The first step in implementing new changes to conventional construction is to understand the barriers or the gaps in adopting new systems. This requires a gathering of various viewpoints from different stakeholders and a good communication model. I had the privilege of doing this exercise to meet with different stakeholders and advise the New Zealand ministry office about my findings.

What the future holds

Particularly in earthquake engineering related to structures, a lot of progress is being made. I cite a few below:

(i) Structural and geotechnical engineering:

Usually, structural and geotechnical engineers do work independently and minimum level of interactions take place before deciding engineering solutions. After the recent earthquakes, there is an increasing trend to change this culture. Geotechnical engineering is gaining more focus.

(ii) Performance of structural elements (skeleton) of the system vs. non-structural elements (fit outs):

There is a growing research interest in understanding, designing, and installation of the fit-outs that support building services. Fit-outs cost much more than the structure itself. Development of new types of structural systems and non-structural systems that sustain minimal damage are emerging.

(iii) Loss estimation of structures due to earthquakes:

This area deals with: what type of damage will occur, what could be the cost of repair, repair time, how long before the building could become unusable and how to make the tenants vacate the building in an expedient manner.

(iv) Real-time monitoring of seismic response of structures to advise stakeholders:

This is to understand the likelihood and extent of damage that can be smartly communicated to the owners or stakeholders to help in making decision regarding 'occupational safety'.

(v) Performance of infrastructure:

Damage potential to various infrastructure networks: underground pipes, cables that interact with soil behavior, roads, electricity and other critical facilities like hospitals, airports, ports. Every stakeholder wants to know what is the risk their portfolio is likely to face.

(vi) Dealing with uncertainties:

Like in every engineering discipline, earthquake engineering has to deal with lots of uncertainties every stage. Acquiring knowledge and skills to deal with uncertainties in various types of modeling would be an added advantage.

Summary

Through my career story you can see that I got what I did not want in the beginning - i.e.. Civil Engineering. And I did not end up doing what I wanted - i.e. teaching. I took risks such as quitting my job before knowing what I would do next and moving to a country I did not know - i.e. New Zealand.

But in spite of it all, I am living in a wonderful family friendly place, working on a topic of national importance. My education fully helped me to get a start in this field. Today my work advises the minister's office on matters related to earthquake engineering.

So what does this mean? It means that one must be willing to accept risks. One must be willing to accept that the outcomes from what we invest in now are not pre-determined. We should be courageous to embrace the new. We must face challenges at home and work with equanimity.

Rather than perfecting a plan, we can tinker and ask ourselves: *"why not try this? What is wrong if I do this? What could come after all? Let me face as it comes."*

John Lennon of the Beatles famously sang, 'life is what happens when you make other plans.' So let's live life in the moment, embrace change!

About the author:
S.R. Uma works for GNS Science, a Crown Research Institute in New Zealand as an Earthquake Engineer. After finishing her bachelor degree in Civil Engineering at PSG College of Technology, she did her post-graduate studies at IIT Chennai. She started her career as an academic and continued to serve in various research organizations. Her research interests include seismic assessment, seismic monitoring of structures and understanding the needs for post-earthquake functioning of built-environment in cities.

Question & Answer

Which leader do you admire the most?

I admire Steve Jobs as a business executive and a leader.

He had at least 3 qualities (not to mention his genius in product design and marketing) a clear vision, passion for the company and its people, and an ability to inspire trust.

There is a passion exhibited by consumers for Apple and for Steve Jobs that is rare in the corporate world.

Natarajan
Mechanical Engineering
Atlanta, USA

INFRASTRUCTURE ENGINEERING

ENGINEERING THE INFRASTRUCTURE FOR FUTURE INDIA

Journeying all over India, the author gives an incisive account of where the infrastructure development needs are and weaves his experiences on what works and what does not.

Introduction

I started my career with the Neyveli Lignite Corporation as a power Engineer soon after graduating college in 1986. However, the drawbacks of public sector, like monotonous work and no opportunities for accelerated individual development drew me to start looking for private sector jobs.

I joined Siemens India in 1995 and worked there for over 11 years. Here I executed projects for various customers and travelled to different parts of the country like West Bengal, Bihar, Haryana and Chhattisgarh. In this process, I came across various cultures and different types of people. Moreover I could see vast difference in development across various parts of India.

I continue to be in power sector today and for the past ten years have been with a major infrastructure group. My observation is that the overall cumulative infrastructure development across India is still not at par with other developing nations. I am happy to note that my current organization has given me a vision for integrated growth prospects for India in the years to come.

According to The World Bank, lack of infrastructure is limiting the growth of the Indian economy. For India to progress from the present state, infrastructure readiness is very important. Infrastructure planning has to be engineered in such a way that it will not be a hindrance for growth in the future also.

In this section, I will describe what the various sectors of infrastructure are, and will outline the type of infrastructure investments needed in the future.

The various infrastructures that need to be geared up in India include Energy, Transportation, Infrastructure and Agriculture

Energy

Within the energy sector, there are sub-sectors such as Power and Oil & Gas

Power

India consumed around 756,000 gigawatt-hours of electricity in fiscal 2012, according to a report titled Energy Statistics 2013. Total demand of Power is expected to increase substantially and accordingly India's electricity demand is expected to grow by 132% by 2035. The government plans the addition of new generation capacities to the tune of 89,000 MW during the period 2012-2017. Considerable investments are also expected in the transmission & distribution network.

Though the current focus is on fossil fired power stations, owing to the gap in demand and supply, the renewable energy sector is also picking up. In the last decade a lot of favorable policy changes brought rapid expansion in this sector, with new power plants coming up and private players entering the market. However during the same period, not much was done to improve the transmission and distribution section and the Transmission & Distribution projects then were mired by right of way issues and lack of governance.

China on the other hand has come out very strong in resolving this issues and the country has witnessed rapid economic growth in the last decade owing to the infrastructure development. The best example would be comparing The Three Georges Dam in China to the Sardar Sarovar Dam on River Narmada in India. Construction on three Georges dam started in late 1994 and completed in mid-2014 and the Sardar Sarovar Dam, which started in 1979, is still not complete in all respects.

To reduce the carbon footprint and moving towards a green energy goal, the Government of India (GoI) is promoting renewable energy (RE). Though the GoI has come up with various incentive policies in this sector, the growth is still not sustainable. A lot needs to be done in terms of framing new policies and financing RE projects to see rapid development in this sector.

Today the government is trying to frame policies in this sector and make up for the lost time in the last decade. The free electricity and subsidies declared for political advantage should be scrapped for better growth of this sector. The quality of the grid to be refurbished to a smart grid to take of care of the mix of generations fluctuations from renewable, fossil and nuclear energy.

Moreover, realistic contractual agreements shall be made while entering to long-term agreements by Government and generating bodies for fuel and purchase of electricity. Stringent enforcement of laws to avoid power thefts should be done. Technological developments should be undertaken to reduce losses at various stages. The methodology for clearing a project to take off should be streamlined and faster.

The manufacturing infrastructure for this power sector has not developed over these 60 years whereas in China it has gone up to many manifolds. This will help India to reduce the dependence on other countries for project stage and maintenance stage. The infrastructure for any sector needs to be engineered in an integrated way.

Main thrust needs to be given for this sector in the following for future:

o Development of manufacturing facilities and allied infrastructure like foundries and workshops
o Quick clearances for project
o More focus on renewable, nuclear
o Grid stability, smart grids , strong transmission & distribution network

Oil & Gas

The rapid industrial growth that is being forecast by year 2020 will give an impetus to the demand of Oil and Natural Gas in India. The growth of Domestic energy and the fertilizer sector is expected to raise the demand of Gas in the current decade. According to the Planning Commission of India, during the Twelfth Plan, the total domestic energy production is expected to reach 670 million tons of oil equivalent (MTOE) by 2016-17 and 844 MTOE by 2021-22. This will meet about 71% and 69% of the total energy demand the balance shall be met by Imports.

The Government of India (GoI) has already started taking steps to encourage the development of alternate fuel sources such as coal bed methane (CBM), gas hydrates, hydrogen fuel cell, and blending of bio-fuels under the Ministry of New and Renewable Energy to reduce the country's dependence on imported fuel. These new technologies, enabling efficient use of fossil fuels working in tandem with renewable energy are expected to fill the demand-supply gap in the future.

Transportation

With a growing population in India, demand for transport would increase further by 2020. Though initiative has already been taken for improvement in this sector, still much needs to be done.

All-weather rural roads are expected to provide access to the farthest outlying villages. Due to lack of roads, some parts of the country where agriculture produce is surplus are unable to connect to the National market. This is to be planned and initiated immediately as it will help the commodity to reach for distribution in time and significantly reduce cost. This will also help in improvement in Health care facilities available to such far-flung Villages.

Some of the roads, bridges are planned and made but due to delay in execution faced problems of cost overrun. Also, due to inadequate planning, after completion they become insufficient for the existing population. The poor quality of roads increases the fuel cost and maintenance cost of vehicles. Its leads to unsafe travel too. The infrastructure facilities such as airports, railway stations and bus depots have recently become short of capacity in handling the increasing traffic.

As increased traffic is expected at Ports, owing to the rapid industrial expansion, they need to be well connected to the inland for fast material movement. Dedicated freight corridors should also to be envisaged for fast movement of goods and material from and to the Ports. According to the Ministry of Railways' estimates, demand for passenger and freight services would surge, which would require expansion of 25,000 km of new lines by 2020. The development of fast inter-city rail services is expected to increase passenger train traffic by 2020.

As India is a land of large Perennial Rivers, the interlinking of rivers is to be implemented and Water transportation should be increased to meet the raising demand. The non-perennial rivers will get continuous water supply due to interconnection and this shall significantly improve traffic and reduce the transportation time.

By improving the construction technology using latest sophisticated equipment and fast construction methods the quality, reliability and maintainability of rail and roads shall be improved. Adequacy has to be engineered keeping into mind Safety and future expansion. Barricaded roads and rails for faster movements need to be conceived and implemented.

Infrastructure

Urban Infrastructure
The Jawaharlal Nehru National Urban Renewal Mission (JNNURM) was launched by the Ministry of Urban Development for a seven-year period (2005-2012) to encourage cities to initiate steps for bringing about improvements in their civic service levels in a phased manner.

This covers urban renewal, water supply (including desalination plants), sanitation, sewerage and solid waste management, urban transport, development of heritage areas, and preservation of water bodies. The most notable aspect of this is the BRTS buses, which have become a major success in most of the cities.

Today though cities are fast developing, there are various pockets where basic amenities like clean drinking water and sanitation are still not developed well. Prioritizing the development of infrastructure of these areas is of the utmost importance.

The Government is considered as the sole financer of infrastructure projects over the years. However, given the priorities of the Government and lack of budget, the financing of infrastructure projects has slowed down in the last decade.

Although, the GoI is still the largest financer of infrastructure projects it is encouraging private investment and looking forward to more Foreign Direct Investment (FDI) in the Infrastructure sector.

As a result, the share of the private sector in infrastructure financing gradually increased from a mere 25% in FY05 to 33% in 2010 and is expected to increase further to 45% by 2020.

Rural Infrastructure
India is a country of villages. However, since independence not much has been done to improve the infrastructure of rural areas. Even today in most of the villages, the clean water, motorable roads and sanitation are distant dreams. Though electricity has reached some villages, the quality and availability is an issue. Many villages in Rajasthan, Bihar are without proper sanitation facilities. Even in many parts of extension of metros the condition is pathetic.

We need to focus on rural areas so as to bring about development in our country. By empowering the villages we shall be able to improve the quality of living of the villagers.

Both public and private initiatives are needed to push the infrastructure development in rural areas.

Social Infrastructure
Health & Sanitation: Improvement in health and sanitation facilities can be achieved through improvement in access to and utilization of health, family welfare and nutrition services with special focus on the under-served and underprivileged population.

Public and Private expenditure on health and sanitation is to be increased with the advent of "*Swaccha Bharat Abhiyan,*" an initiative of the GoI. Furthermore, health insurance plans as well as government schemes such as 'National Health Insurance Scheme' for socially vulnerable and low-income people are also expected to play a role in financing for quality health services.

Education: Though literacy rates in India have increased considerably, from 18% in 1951 to 65% in 2001 and 74% in 2011, they vary substantially among males and females as well as urban and rural populations. With the projected increase in the per capita income and various initiatives of the GoI, promoting education or females especially, a high enrollment ratio is expected in this decade. This in turn will up the literacy rates further by 2020.

However, not every private educational institute has the quality of a Birla Institute of Technology or our own PSG College of Technology. The private educational institutes founded aplenty in recent year, that have come up at a very large scale are only concerned about making money, and they are running the institutes as business function rather than imparting skills to the students. Quality of education is very poor and this has resulted in large number of unemployed individuals after college. The government has to intervene and make education affordable as well as skill oriented.

New universities and institutes at the central, state and local level by both Government and private players imparting quality education will give a boost to the education infrastructure of India.

Agriculture
With the growing population the demand for food grains is set to increase further. Keeping pace with this demand the food grain storage, handling and classification system has to be developed.

With the Government of India (GoI) keen on developing new technologies in seed germination, irrigation, harvest, storage and handling and the backing of various institutions and industries this sector shall witness a tremendous growth.

Owing to lack of infrastructure most of the food grains are stored in the open resulting in wastage and also damage due to weather. This in addition to lack of transportation facilities from warehouses to markets leads to inflation.

For non-perishable items, the post-harvest loss accounts for 5-10% and for perishables it is about 30%. New initiative will be required by both Public and Private Players to bring down wastages. The whole new system from harvest to plate has to be stepped up and new infrastructure in this sector will have to be developed on war footing.

Irrigation
Though Monsoon always tries to play a truant - an increase in irrigation facilities will increase the reliability in this sector. Though the schemes of irrigation are largely funded by the public, private players will always play an indirect role in its development.

The central Government has already declared various irrigation projects as national projects. These projects are expected to irrigate about 2.1 million hectare of farmland apart from engendering additional indirect benefits and availability of drinking water. Primarily funded by the GoI, these projects will cater to the food needs of the country by 2020. As significant investment is expected in irrigation facilities, timely and effective implementation of irrigation projects coupled with good governance will ensure achieving the desired results.

However, construction of dams takes a lot of time. The government should focus on constructing anicuts. Construction time of anicuts is small and also there is no submergence issue. Micro irrigation systems such as drip irrigation are to be implemented in all water-scarce and rain-fed areas. With the lowering of water table in most part of the country the investment in digging a well is increasing substantially. Practices like rainwater harvesting will have to be practiced religiously in both urban and rural areas to recharge the water table.

Technology and research & development
In addition to development of irrigation facilities, increased investment in technology, especially information and communication technology (ICT) will help drive agriculture growth.

The satellites provide accurate data for the farmers to plan the sowing. High-grade seeds shall be developed to increase the yield. Technology is going to play an important role in the agriculture growth of this country in the future.

Cold storage and post-harvest management
Post-harvest losses in India are currently 5-10% for non-perishables and about 30% for perishable farm output. In view of this, greater focus is on development of post-harvest handling and agro processing during the current decade. The capital investment subsidy scheme implemented by the Ministry of Agriculture is also expected to support investments in cold storage and rural godowns. While public investments are expected to provide the much-needed support for development of post-harvest infrastructure, special thrust will be needed on encouraging private investments in this segment. Subsidy under National Horticultural Mission and the scheme for development and strengthening of Agricultural Marketing Infrastructure, Grading & Standardization are likely to attract private investments.

There are other miscellaneous infrastructure investments that will help accelerate agriculture sector growth directly or indirectly which include: FDI in retail chain, Infrastructure development, especially transportation and increased supply of power, Development of Mega Food Parks, Development of Horticulture, Precision Farming Promotion, Early Warning System and Weather Watch Management, Agricultural marketing promotion.

Agricultural reform

Policies coupled with infrastructure drive the Economic growth. GoI has devised several policies to promote agricultural growth in the past but further work needs to be done. Expecting a high growth in this sector in the current decade, the preparedness shall have to be ensured. Some major policy initiatives that need to be implemented in the current decade are:

- Agriculture policies
- Industrial policies
- Fiscal Policies – Taxation and subsidies
- Financial policies

- Environmental policies

Recommendations:
- All projects must be engineered with forward and backward integration
- All projects are to be engineered considering provisions for future expansion
- Government clearances for the infrastructure projects to be speeded up
- Quality Education rather than quantity education
- Implementation of new technologies in agricultural sector
- Implementation of interlinking of rivers and planning for anicuts
- Proper planning, budgeting and monitoring for schedule completion

Summary

I have tried to provide here a broad overview based on my experience in the state of infrastructure in India as well as the areas that need focus to move India to a 21st Century economy. As our current Prime Minister Narendra Modi has repeatedly emphasized, nearly a third of our population is under 30. Thus, the future growth prospects of India are very strong. However, as I have detailed above, this growth can become sustainable nation building only if it is done in a planned and integrated way. The growth of engineers and infrastructure development go hand in hand.

I am very hopeful that the current graduating class and the ones coming up in the future will focus on building their careers in the infrastructure space and contribute the development of this most vital aspect of our economy.

About the author:
Ravishankar started his career with Neyveli Lignite Corporation. He joined Siemens India and took on assignments to extend life of Power Plants. He joined PPIL (A joint venture between Siemens AG and BHEL) in 1998 and contributed in successfully restarting generation in the Thermal Power Plants located in various parts of India. After a short stint in Reliance Energy Limited, in 2006, he joined Adani Group, where he worked on commissioning India's first supercritical unit and also the largest single location Power Station.

Question & Answer

Which leader do you admire the most?

Mother Teresa was an exceptional leader who proved money is not the criteria to become a leader. She started her service to the needy with just Rs. 5 in hand.

She had the ability to build a team and to inspire her team to transcend their self-interest, to attain the highest-level of performance.

Sumathi Singaravadivel
Electrical and Electronics Engineering
Thanjavur, TN

OIL & GAS EXPLORATION

Rigging for Energy

Journeying all over the world and adventuring to oilrigs far off the coast, the author reports back from the frontier. He explores what new fields of study are being developed for the next generation of adventurers in oil exploration.

Introduction

We live in an energy-craving world. Mankind needs energy to power the cars, buses, trams, air-conditioners and everything else that requires to be plugged into an energy source.

While wind and solar energy have made spectacular inroads into how we harvest energy, conventional fossil fuel continues and is expected to drive the market for energy needs over the next three to four decades at least.

The demand for more energy is pushing the frontiers of Oil exploration to new depths.

Pursuing a career in Energy Engineering

I take this opportunity to present to you two post graduate courses in engineering, which is gaining prominence. These courses can be taken by any branch be they civil, mechanical or electrical engineering.

The courses are Master's Degree in *Subsea Engineering* and Master's Degree in *Arctic Engineering*. Let me briefly outline these courses for you:

The familiar 'offshore rigs' is a phrase now turned on its head. Rigs can be seen hundreds of kilometers off the coast, if not more these days.

As the rigs move farther offshore and deeper, the engineering challenges that come with it are immense. From sheer water pressure to corrosion to the challenges in installing and operating equipment & pipelines on the seabed call for specialized 'design, build, operate and maintain' skills.

Subsea Engineering:

This is where a Master's Degree in Subsea Engineering helps.

From entry-level engineers in all of the above design, build, operate and maintain phases, this master's degree is well sought these days in the energy sector. A few universities in the US offer this course.

But, Norway has the reputation of being the better option. Courses are in English language, which obviously can be quite helpful to those of us from the Indian subcontinent!

On successful completion of the course, an entry level start of US$ 350~400 per day can be expected with an international oil field developer or operator.

Arctic Engineering

The Master's Degree in Arctic Engineering is again a two-year course that not only relates to Oil & Gas sector but also broadly outlines 'living, working in and protecting the Arctic Environment'.

Obviously the effects of ice and the extreme sub-zero temperatures are the challenges here. Coupled with the extremely fragile eco-system in the arctic, any disaster in a man-made installation – say a north or South Pole expedition base to an oil rig or an underwater research on sea lions and the like can have disastrous consequences.

Pouring concrete with immersion heaters built into the steel work and powering them up after the pour is one way concrete is made to set in these treacherous climes.

How do you power cranes and forklifts in an arctic environment? Battery powered options become a necessity. Even special grades of diesel don't fire on a cold wintry morning.

Challenges and demands

A diet of more than 9,000 calories a day is vital to survive if you work in the open – exposed to the elements for even 2 hours a day.

Oil companies, research agencies, environmental groups, think tanks, defense equipment manufacturers, all recruit Arctic Engineering Master's degree holders.

These master's degree courses are predominantly in Alaska, US and Norway. Though civil or marine engineering graduates prefer this course, it opens great vistas in careers for all discipline graduates.

Salaries are rewarding and the career is very challenging.

It is cold though!

Summary

As I have narrated here, after getting a start in Mechanical Engineering, my career and business travels have taken me far and wide - indeed as far as onto oilrigs. It has been both an adventurous and satisfying life.

There is plenty of innovation taking place in energy engineering, as I have outlined above. Mankind has always had a frontier spirit. We are seeking farther and farther limits to learn more about the world and universe we live in and also to find source of value for humanity. Energy is one such search we are constantly engaged in.

For those that choose this path, exciting discoveries await!

About the author:
Deepak Edwin works in the energy industry in Europe. He graduated with a degree in Mechanical Engineering from PSG College of Technology.

Question & Answer

Which leader do you admire the most?

There are many leaders to admire and different reasons for choosing them.

In particular, I admire Kamarajar for simplicity, Rajiv Gandhi for broad thinking and Indira Gandhi for her boldness.

H.K. Siva
Textile Engineering
Hong Kong

SOUND ENGINEERING

The Sound of Music

The author takes his passion for music from an early age and converts it into a pursuit for his family and community by also utilizing his engineering training.

Introduction

Being a former member of my college music group, appropriately called "TechMusic," the mention of PSG College of Technology always brings back golden memories of old melodies to my mind.

In addition to regular concerts on campus, TechMusic also used to give public performances off-campus for the community in and around the City of Coimbatore. Once, TechMusic had to perform on campus soon after well received open air performances at the hill station of Nilgiris and at the nearby town of Thirupur.

The band was confident that the performance at the PSG Assembly Hall would be a definite success following our hit performances off campus. To our disappointment, the show at our college campus was a disaster despite excellent individual performances from each member of the team.

In both the open air performances out of town, professionals were hired to take care of the audio set up. However, for the concert in our own assembly hall, we missed the extensive preparatory work that goes into setting up the sound system. Without the perfect blend of the vocals and instruments, harmonious music was simply missing. With an openly critical review from our college community on this audio fiasco, we were careful not to repeat the mistake again in the future. For me, this lesson went beyond this searing incident on the college campus amidst my friends and professors.

Resonating Reconnection

My music activities in terms of my own performance took a backseat when I started a full time job following my graduate studies in computer science and business. Years later, I felt reconnected when my kids started getting involved in music. Attending my son's Jazz and Carnatic music concerts rekindled my appreciation towards the art and technology of media production, delivery, and the electro acoustics involved in sound engineering.

31

Early this year, we got an opportunity to experience professional music production in a local recording studio designed by a world renowned studio designer. Vocalists, percussion and instrumentalists assembled to produce high quality recording with the guidance of a Carnatic music maestro from Chennai. This fellowship project was organized by a commercial venture founded by a young entrepreneur with an engineering degree from IIT and MBA from the MIT Sloan School.

A few months later, my son played the percussion arrangements for a dance production. This sold out show performed entirely by young students, involved live and recorded music. For producing the recorded music, my son worked with a team involving engineering, business and medical students. The team using entry level recording software and hardware was able to produce inspirational music of studio quality. Such ease of production, enabled by technology combined with the powerful reach of the internet, made it possible for Boston based Shankar Tucker to make it big in Indian music through his *Shruti Box* recordings by using a "Viral Video" approach on YouTube.

Innovations in Sound

Human beings are very sensitive to sound quality. A person may find it hard to guess the correct scale of a music piece, but can instantly recognize a sour note in the rendition. It may also be hard for people to compromise on sound quality.

Disappointed with the inferior quality of a stereo that he bought, Dr. Amar Bose from the MIT invented a new type of stereo speaker in the early 60's. Later, he founded Bose Corporation which became synonymous with high-quality audio systems.

Later, with a combination of iTunes and the iPod, the great innovator, Steve Jobs catalyzed personal music consumption and also set the stage for launching the iPhone in Apple's revolutionary comeback as market leader. Apple's investment in music continued with their recent acquisition of Beats Electronics, a company known for its premium audio products and services.

Music holds a special place within our hearts and provides opportunities for engineers and artists to blend technology and art. Designing and delivering quality audio products and services involves multiple fields of engineering.

Sound and Sound Engineering

Sound is a longitudinal wave which travels outward in a medium. It is slower than light and is subject to reflection, refraction and interference. The lifecycle of sound involves the Production, propagation and perception of sound, also known as the three P's of sound. Audio engineering involves two broad categories of specialization – Sound Systems Engineering and Product Research and Development Engineering.

Sound System Engineers are the brain behind the audio system setup for successful performance shows and sound production. They understand the requirement of the show or production at hand. Based on the available products and budget, they design a solution with their technical expertise and field experience as studio engineers and sound designers. The challenges with large live performances for both indoor and outdoor are numerous. It takes a team of seasoned engineers and media experts to bring success to big shows.

Product research and development in sound engineering includes architectural acoustics that deals with engineering sound quality in a performance hall through sound reinforcement. It also involves electro acoustics dealing with the design of products like microphones, headphones, loudspeakers, recording technologies and sound reproduction systems. With strong technical background and solid musical ideas, the audio engineers strive to create sonic signatures and brands such as Allen Heath, Shure through innovation.

New frontiers in Sound

From vinyl records to virtual reality, a revolution is taking place in the field of audio engineering, from personal space to performance stage, with the decibels of pleasure peaking decade after decade. Movie production and gaming industry are driving technologies like spherical audio for total audio immersion. 3D headphones are already in the market, setting the trend for the next decade.

Princeton University has developed a way to play three dimensional sound recordings from standard stereo laptop speakers. 3D Soundscape technology has crossed the boundaries of entertainment space into human wearable application. The 3D wearable camera for the blind from Microsoft, tricks the wearer's brain into thinking they come from certain directions creating a "3D soundscape" for safe navigation.

Resources for acquiring knowledge

Audio engineering can be a highly rewarding career or a favorite hobby for an engineering student with a love for music. Convergence of computer technology and Music Instrument Digital Interface (MIDI) has ignited a revolution in recording and music production industry.

A.R. Rahman's KM Music Conservatory in Chennai in collaboration with Middlesex University in the UK offers audio engineering and electronic music production courses. Carnegie Institute of Technology in the US offers a Master of Science in Music and Technology. The program offered both by the School of Music and School of Computer Science consists of courses spanning both music and technology and a comprehensive capstone project.

There are many magazines and journals in print and electronic media making the learning of audio engineering a fun experience. Magazines like "Sound on Sound" recognized as the bible of hi-tech music recording industry, publishes highly informative articles on music technology, and also features in-depth product tests of monitors, microphones, mixing consoles, keyboards, synthesizers, music computers, and virtual instruments.

The Audio Engineering Society (AES), a professional society devoted exclusively to audio technology founded in the United States in 1948, has grown to become an international organization today uniting audio engineers, scientists, students and creative artists worldwide. The society helps in spreading knowledge in research and development in the field of audio engineering.

AES conducts a worldwide Student Design Competition during AES Conventions as an opportunity for aspiring hardware and software engineers. Students seeking bachelor, master or doctoral degrees participate in the audio product design competition and get recognized for their ingenuity, technical creativity, and hard work.

Summary

My interest in audio engineering and technology was rekindled by my involvement with the music activities of my kids. Even with little exploring, I realized that the field of audio engineering is so extensive, yet exciting. Scientists and engineers through innovation have brought powerful applications to the homes of musicians and listeners to create or enjoy wonderful music. As technology pushes the boundaries in this domain every day, the future world of the sound of music sounds exciting and offers glorious career opportunities for the "Tech & Music" oriented talent.

About the author:
Nambi Thirumalai lives with his family in Dallas. He is an Information Technology professional. He was a day scholar at PSG, graduating in Electrical and Electronics Engineering.

Question & Answer

Which leader do you admire the most?

As a person, I most respect Kanchi Sri Chandrasekharendra Sankarachariyar, for his simplicity, his intelligence, living by what he preached.

Gauthaman
Civil Engineering
Tiruchinapalli, TN

TEXTILE ENGINEERING

Fiber Materials for a Better World

Of all the technologies we consume, textiles are the least intrusive from a technology perspective and the most intrusive from a fashion perspective. The author of this chapter traces the fundamental role that textiles has played in the industrial revolution and points to the exciting developments in the future for this sector.

Introduction

I was born and raised in Coimbatore. My education through high school was within the PSG Educational environment. I grew up watching the pavilion that used to be in front of the PSG College of Technology building with a globe on the top. That pavilion was created as an exhibit at a world engineering fair. It has long since been taken down. As a kid, I attended every exhibition hosted at PSG College of Technology. When it was time to go to college, there was no question where I wanted to go.

My father was in textile business. Most of my relatives and friends were in the textile business. When I was growing up, on my way to school and back, I had to ride my bicycle past four textile mills. Thus it was another easy decision for me in terms of picking which branch of engineering I wanted to pursue.

Looking back, little did I now that the classes in material science, mechanics of machines and applied mechanics I took when I was in college, would be just as impactful in shaping my career as the courses in textile engineering.

Onwards to the United States

I had the opportunity to study nonwovens and polymeric materials at the University of Tennessee. I have since spent my entire career as an R&D Technologist in modifying fiber surfaces and participate in various ways fiber materials make this world a safer and a more comfortable place to live.

As I embarked on my career, I was amazed at the indispensable value of textiles in filtration, noise control and bio-barriers. Tremendous surface area of the extremely lightweight materials improve the quality of air at home and reduce the emissions from the industries as filters, create a quiet ambience at homes and automobiles as barrier fabrics and protect the patients and healthcare providers as surgical gowns. Innovation in the textile material science is at its best when the textile engineers collaborate with biologists and engineers from other disciplines.

A long way from Luddites

The textile industry was at the birth of the industrial revolution in the 1800s in the United Kingdom. The power loom reduced demand for skilled hand weavers, initially causing reduced wages and unemployment. Protests followed its introduction. In 1816 two thousand rioting weavers tried to destroy power loom mills and stoned the workers. In our high school reading assignment, one of the stories we were asked to read a story that was about the innovator who invented the *spinning jenney*. His family in the story was under threat of workers who were losing their jobs. As a young student, I did not appreciate the forces of disruption unleashed by innovation. But the emotions contained within the story have stuck with me all these years.

The *Luddites* were a social movement of British textile artisans in the nineteenth century who protested (often by destroying mechanized looms) against the changes produced by the Industrial Revolution, which they felt was leaving them without work and changing their way of life.

The movement emerged in the harsh economic climate of the Napoleonic Wars and difficult working conditions in the new textile factories. The principal objection of the Luddites was to the introduction of new wide-framed automated looms that could be operated by cheap, relatively unskilled labor, resulting in the loss of jobs for many skilled textile workers.

In modern usage, *Luddite* is a term describing those opposed to industrialization, automation, computerization or new technologies in general.

In the longer term, by making cloth more affordable the power loom increased demand and stimulated exports, causing a growth in industrial employment, albeit low-paid. The power loom also opened up opportunities for women mill workers.

This disruption visited the United States in the 20th century as well. The Carolinas, where I work, were once the epicenter of the U.S. textile industry, but since the late 1990s, thousands of jobs were lost when emerging markets joined the game, touting cheaper materials and labor. Carolinas textile jobs went to China, Brazil and Vietnam, among other places. But now, in an ironic turn of events, Chinese companies are looking to manufacture in the U.S., lured by lower costs of energy, cotton and land,

and wary of rising labor costs in China.

This is then a key lesson of innovation in the capital markets oriented economies. *Innovation causes enormous disruption and pain.* But it also opens up great new opportunities in both technological and cultural arenas. And in turn, that sets up societies for the next stage in their development.

New advances - Synthetic Fibers

Synthetic fibers are lighter than metals but significantly stronger. The high strength to linear density ratio and flexibility of the polymeric fibers allow the fiber materials to make a huge difference in the everyday quality of life. Composite fabrics from *Kevlar* to *Nomex* can not only stop the bullets in war zones and protect the firemen in wildfires, but protect fingers and arms as cutting gloves in food processing. Areas of application can be as wide as improving air quality, reducing pollution, protecting from disease causing germs, reducing energy consumption, preventing soil erosion, protecting water resources and last but not least, more comfortable fabrics.

The US Department of Energy expects that the carbon fiber reinforced composites can lower the weight of an automobile by 50% and reduce energy consumption by 35%. Similar fiber based composites make the blades in the windmills and wings on an airplane lighter and stronger. Studies have shown that high performance apparels can not only reduce muscle vibrations in athletes but improve blood circulation in the diabetic patients. Three dimensional fabrics find home as stents in the arteries to the heart as well as support structures in the abdomen after a hernia.

Can we imagine an automobile without seat belts or an airbag? Tires last longer and run faster thanks to tire cord fabrics made from high modulus and low shrinkage yarns. Textile materials protect the drivers while keeping the weight of automobile and fuel consumption lower.

Wearables and Textiles

Another area where these is enormous progress is in the area of wearable technologies. When we mention wearables, we immediately think of gadgets on our wrists. However, a bigger revolution is taking place with wearables within clothing.

Sri Lanka is a good example of a country making bold strides in this regard. Sri Lanka has the highest per capita apparel exports in the Asian region. Their garment manufacturers are adopting state of art technologies involving wearable electronics, e-textiles and smart clothing for the global apparel market and the use of environment friendly fabric treatment and color processing ingredients.

The application of this innovation for providing our societies with data about our health and wellbeing and creating programs to promote physical and mental fitness is enormous.

Textiles for the 21st Century

Our impression of textiles as a study of making clothing is far from the reality in 21st century. The challenge to new textile graduates is in imagining the places and things that fibers can make a paradigm change in India and other developing countries.

Textile materials afford the new graduates wide opportunities to do good for the society while embarking on successful careers in developing innovative solutions. There are fiber materials that are biodegradable in the landfill such as polylactic acid - imagine carry bags that can be used forever and turn biodegradable after disposal or capable of transporting data signals across the continents in optical fibers. Carbon fibers are much lighter than aluminum but much stronger than steel alloys.

Summary

Living in the United States for the past 30 years and working with clients globally has afforded me the opportunity to witness first hand, the rapid changes in textile technology. My key takeaway is that almost all the successful innovations and developments have at least one thing in common: *they positively impact the human living condition*; whether it is personal safety or comfort or the environment. As the new graduates enter the profession, whatever function they perform, product they develop or produce, or any problem they solve, they must remain cognizant of the role they play in making this world a better place.

About the author:
S. Ranganathan graduated from Sarvajana Higher Secondary School. After graduating from PSG College of Technology with a degree in Textile Engineering, he received his M.S from the University of Tennessee and an MBA from the University of North Carolina at Charlotte. He joined Goulston Technologies, Inc. after graduation and has remained there since. He currently serves as Executive Vice President and Chief Technical Officer.

Question & Answer

In the context of online learning, what is the redefined role of a teacher?

President Abdul Kalam's answer to this same question in a teacher's meet at one of the schools in Dubai, which I had the opportunity to attend, was that the teacher and student interaction would never vanish; and that this interaction would have to take place in a classroom.

Even if there are plenty of Massive Open Online Courses (MOOCs) available, the easy and simple way to understand a concept is by interacting with a good teacher.

Be proud of a teacher!

Sriram Balakrishnan
Electrical and Electronics Engineering
Dubai, UAE

ELECTRONICS AND COMMUNICATIONS ENGINEERING

Advances in Communications and Signal Processing

As all of us experience everyday, we are in an increasingly wired world. To state it better, we are in an increasingly wireless world. Undergirding that is a rapid development in signal processing and communications technology. The author traces his career in this critical area and offers his view on what is at the forefront of these developments.

Introduction:

Electronics and Communications Engineering was a new and emerging branch of study when I started my undergraduate program at PSG College of Technology. I am thankful to Prof. Jayaraman and Prof. Narayanaswamy for triggering my initial interest in the field of electronic communications.

It was also my dream to acquire a masters degree from the Indian Institute of Technology (IIT). However my entrance exam score (GATE) was not good enough to gain admission. Rather than take up a job, I was determined to pursue my interest. With the support of my parents, I decided to stay at home and prepare for my examinations once again.

I succeeded the second time and I joined Master's program at the Indian Institute of Technology in Chennai, specializing in Communication Systems.

ESP on DSP

The IIT curriculum was well designed and quite advanced, especially in the field of Digital Communications and Signal Processing.

The field of DSP (Digital Signal Processing) got more attention with the advent of DSPs(Digital Signal Processors) and it was around 1983 when the first DSP (Digital Signal Processor) from Texas Instruments was introduced in the market.

I got the opportunity to design and implement a low bit-rate (2.4 kbps) MODEM (Modulator/Demodulator) using the Texas Instrument DSP. This experience triggered lots of interest to work in core signal processing and communication projects in my career.

A Career in Signal Processing

My first job was with Indian Telephone Industries in Transmission Research & Development. The project was to design and develop an end to end secure communication systems for the Indian Army. This project involved designing hardware and developing algorithms and DSP software for subsystems such as speech coders, encryption engine and wireline MODEMs (Modulator/Demodulators) using DSPs. This was one of the more complex projects that I had worked on.

Later on at Motorola in Bangalore, I worked on multiple software projects in DSP and Networking and the focus was to develop well documented 6 Sigma quality software adhering to CMMI policies. I had the opportunity to implement V.32 bis(9.6/14.4 Kbps) wireline MODEMs using a well-defined software process.

Of course, these days we get data rates up to 100 Mbps through DSL technologies and up to 1 Gbps using wireless technologies.

The algorithms and implementations to realize this was exciting for me.

A signal from the United States

ArrayComm was started by Stanford University graduates. One of the co-founders is Martin Cooper who is the father of cell phones. I got a job at ArrayComm in Silicon Valley, California, through a friend.

ArrayComm was founded to develop *smart antenna* based WLL (Wireless Local Loop) and long range wireless internet technologies. The smart antennas refer to the usage of multiple antennas to transmit and receive signal to improve the capacity and range of the wireless system.

A simple analogy to smart antennas is how the brain combines the acoustic signal received through two ears, the smart antennas employ multiple antennas to combine the received signals for improving the signal strength. At the same time the smart antennas help to transmit the signal in a known direction. The smart antenna technology enables to communicate to multiple users who are spatially separated but in the same frequency band.

Technology well ahead of time

Though Array Comm's vision was great and developed a great technology, it was very futuristic and well ahead of the market demands for increased

43

data rates. The market did not require the spectral efficiency during the decade of the 2000s. Whereas the market necessitates high spectral efficiency and self-adapting wireless technologies in the decade of 2010.

It is important for a technology start-up to track the market requirements few years in advance for it to be successful, not way too much in advance.

For example, OFDM (Orthogonal Frequency Division Multiplexing) technology which is the defacto standard for implementing broad-band wireless and wireline communication systems today was invented in mid-1960s. But no one used the technology until recently due to lack of fast processors to implement the technology and there was no requirement for massive broadband communications in 1960s.

Wireless technology for the future

I now work for a wireless start up called *Tarana Wireless*. The company is similar to ArrayComm in developing advanced broadband wireless communication systems with multiple antennas. The systems can be used as wireless backhaul (think of a wireless replacement for ethernet cable or a fiber optic cable) or the system can be scaled to BWA (Broadband Wireless internet Access) systems, and can be used for Wi-Fi systems in aircraft to communicate to the ground.

Wireless technologies evolve in terms of self-adapting the frequency, protocol, and networking parameters to provide a desired quality of service(QoS). There is lot of ongoing research in cognitive radios which is essentially cloud based dynamic spectrum management and enabling coexistence of different wireless devices with different protocols optimally in a given geographical area.

Cognitive radio and future wireless technology triggers research areas of:

1. Perfecting RFIC (Radio Frequency Integrated Circuit) design to operate in a wide range of frequencies

2. Minimizing the distortion of PAs (Power Amplifier) using DPDs (Digital Pre-Distortion), minimizing the variation of RF device parameters over a wide range of temperature

3. Sophisticated calibration procedures to compensate for the RF imperfections, high precision ADCs that can work at very high sampling rate

4. Configurable processors integrated with hardware accelerators and a well-structured hardware/software module development in an automated fashion abstracting the interfaces between software and hardware

5. Low cost, high density, and faster FPGAs (Field Programmable Gate Arrays) for quick prototype and product development

6. Improvement of material science to perfect the antenna design

7. Cross-layer optimization between PHY (Physical) and MAC (Media Access Control) to increase the spectral efficiency further

8. Predictive channel models using neural networks to predict the wireless channel and optimally transmit data

9. Cloud based network nodes for monitoring and optimizing the performance globally

We are rapidly moving to becoming digital societies across the world. We are moving from being voice focused to data focused. We are moving from building pipes to providing content. We are hurtling from connecting people to connecting machines, through the *internet of things*.

Most companies will be using Cloud-based services. Cooper's law states that the number of voice/data connections has doubled every 2.5 years which means that the average data requirement is going to be 3.2 GB in 2020, which necessitates a global dynamic spectrum management and high spectral efficiency wireless technologies to meet the data requirements.

Summary

Being in start-ups all through my career, I have learned that it is challenging to continuously learn and keep abreast of new technologies. It is very important to learn continuously in the job as the technologies are rapidly evolving and the tools and infrastructure to develop these new technologies are rapidly changing.

The world always asks, especially in Silicon Valley, "*What can you do today?*" and it does not care about what were your previous achievements, accolades, and fame. So, everyone has to be prepared to work on the latest developments as they rapidly evolve. Keeping pace with these rapidly changing technologies is not just about acquiring and keeping pace with new technical skills. It is also having a good mindset and a healthy body to sustain the stamina to stay current and relevant in today's global and competitive marketplace

About the author:
Kamaraj works for Tarana Wireless as Executive Director, Software. He lives with his family in Sunnyvale, CA.

Question & Answer

What specific 'green' initiatives are you involved in?

In our plant in Dubai, we took this initiative:

 We have large number of window and split Air Conditioning units installed. During summer months when the humidity hovers in their 80s, these units discharge condensate water continuously. So we tapped all the condensate water and diverted them to water plants we have nearby using a piping system. It turns out that this meet the complete water requirement of these plants. It also acted like drip irrigation providing optimum water for these plants to thrive in the harsh summer months.

In addition, our plant's energy conservation cell has taken many projects, which has resulted in power saving of more than 20% with respect to the designed energy requirements.

Let me state that this is an ever-moving target for all of us. So every creative step we take, big or small and whatever we save will really help the future generations to come.

Velayuthan
Production Engineering
Dubai, UAE

ENGINEERING SCIENCE

Challenges for an Engineering Scientist

As a concluding chapter to this section of the book on 'An Engineers Life', the author introduces us to the concept of engineering science and describes how this field can be crucial in many new and cutting edge areas relevant for the 21st century.

Introduction

I got my start training in Civil Engineering when I joined PSG College of Technology. Finite Element Analysis and Boundary Element Analysis were cutting edge concepts in those days, taught by Prof. Rajasekaran. We devoured these subjects with great interest. This inspired me to pursue an advanced career in Civil Engineering. I decided to pursue a doctoral degree in Faculty of Engineering at Indian Institute of Science and was honored to receive a gold medal.

My first role was as a *fellow* at the CSIR-Structural Engineering Research Centre
(CSIR-SERC), Chennai in 1988, where I worked as Scientist at various grades and until the present day where I am a Senior Principal Scientist.

My field of research interests include advanced finite element analysis, computational structural dynamics, fatigue and fracture analysis, steel structures - analysis, design and testing and material science. Through the course of my career I have had the opportunity to publish well over a 100 papers and reports.

I bring all this up to say that you can of course pursue a career in the field and around the world. These examples are well document by my classmates in the preceding chapters.

I would like to introduce you to the life of research and science. Can one be an engineer and a scientist at the same time? If research is your appetite, then engineering science is a terrific path to pursue. So, in the following few pages, I will describe what the field of engineering science is and what kind of challenges and advances one can drive in the 21st century.

What is engineering science?

Engineering science is a multidisciplinary field that involves the practical application of principles and facts dealing with various aspects of our

48

physical or material world. The discipline of engineering is extremely broad and encompasses a range of more specialized fields of engineering, each with a more specific emphasis on particular areas of applied science, technology and types of application. Citing Arvid Eide et al, engineering science involves creative application of scientific principles (i) to design or develop structures, machines, apparatus or manufacturing processes (ii) for utilizing them singly or in combination for an intended purpose (iii) to construct or operate these with full cognizance of their design and (iv) to forecast their behavior under specific operating conditions, encapsulating the economics of operation or safety to life and property in all spheres.

As M.A. Robinson has written in the *Journal of the American Society for Information Science & Technology*, engineers generally make use of their knowledge of science, mathematics, logic, economics and appropriate experience or tacit knowledge and all other available information to find suitable solutions to a problem within the given boundaries and assumptions. Although engineering solutions make use of scientific principles, engineers also take into account of safety, efficiency, economy, reliability and constructability or ease of fabrication as well as the environment, ethical and legal considerations such as patent infringement or liability in the case of failure of the solution. Creating an appropriate mathematical model of a problem allows them to analyze it and to arrive at potential solutions. Usually multiple reasonable solutions exist, so engineers must evaluate the different design choices on their merits and choose the solution that best meets their requirements.

Engineers Can Make a World of Difference

The role played, and the challenges faced by an *Engineering Scientist* is quite different from a Normal Scientist. While, normal scientists try to understand nature to evolve new things, engineering scientist try to make things that do not exist in nature. Engineering scientist stress on invention of something. To embody an invention, the engineering scientist puts his idea in concrete terms and designs something that people can use. That something can be a device, a gadget, a material, a method, a computing program, an innovative experiment, a new solution to a problem or an improvement on what is already existing. Since a design has to be in concrete terms, it must have its geometry, dimensions and characteristic numbers, so that the intended item can be visualized and subsequently productized. Almost all engineering scientists working on innovative solutions find that they do not have all the required information. Most often, they are limited by insufficient basic scientific knowledge about the natural processes or phenomena, which demands perfect understanding about mathematics, physics, chemistry, biology and mechanics.

As Walter Vincenti asserts in his book *What Engineers Know and How They Know It: Analytical Studies from Aeronautical History*, engineering research has a character different from that of scientific research. First, it often deals with areas in which the basic physics or chemistry are well understood, but the problems themselves are too complex to solve in an exact manner.
    "Engineering Scientists always look for cutting-edge solutions, technologies and products.'

Grand Engineering Challenges for Engineering Scientists

From urban centers to remote corners of earth, the depths of the oceans to space, humanity has always sought to transcend barriers, overcome challenges and create opportunities that improve life in our part of the universe. During the 20th century, many *great engineering achievements* have become so commonplace that we now take them mostly for granted. Technology involving *Green and White Revolutions* allows an abundant supply of food including milk and safe drinking water for much of the world. The electricity and electrical and electronic equipment and gadgets have become essential part of many of our daily activities. The travel across the globe has become much easier and affordable enabling movement of people, goods and services wherever they are needed. Growing computer and communications technologies are opening up vast stores of knowledge and entertainment. As remarkable as these engineering achievements are, certainly many more great challenges and opportunities remain to be realized.

Some of the exciting grand challenges are spelled out below, though not exhaustive:

Reverse-Engineer the Brain
A lot of research has been focused on creating thinking machines for e.g. computers capable of emulating human intelligence. However, reverse-engineering the brain could have multiple impacts that go far beyond artificial intelligence and will promise great advances in health care, manufacturing and communication.

Advanced Health Informatics
As computers are available for all aspects of human endeavors, there is now a consensus that a systematic approach to health informatics - acquisition, management and use of information in health care can enhance the quality

and efficiency of medical systems and the response to widespread public health emergencies.

Engineer Better Medicines
Engineering scientists can enable the development of new systems to use genetic information, sense small changes in the body, assess new drugs and deliver vaccines to provide health care directly tailored to each person.

Make Solar Energy Economical
Currently, solar energy provides less than 1 percent of the world's total energy, but it has the potential to provide much & much more. Exciting technological solutions needs to be explored to tap this chief source of energy.

Provide Energy from Fusion
Fusion is the energy source for the sun. The challenges facing the engineering scientific community are to find ways to scale up the fusion process to commercial proportions in an efficient, economical and environmentally benign way.

Enhance Virtual Reality
Within many specialized fields, from psychiatry to education, virtual reality is becoming a powerful new tool for training practitioners and treating patients, in addition to its growing use in various forms of manufacturing as well as entertainment industries.

Secure Cyberspace
Computer systems are involved in the management of almost all areas of our lives, from electronic communications and data systems to controlling traffic lights to routing airplanes. It is clear that engineering scientists need to develop innovations for addressing a long list of cyber-security priorities.

Advance Personalized Learning
A growing appreciation of individual preferences and aptitudes has led toward more "personalized learning," in which instruction is tailored to a student's individual needs. Given the diversity of individual preferences and the complexity of each human brain, developing teaching methods that optimize learning will require engineering solutions of the future.

Restore and Improve Urban Infrastructure
Infrastructure is combination of fundamental systems that support a community, region or country. Society faces the formidable challenge of modernizing the engineered structures that will support our civilization in

centuries ahead.

Engineer Sustainable Materials
Future world needs will require materials that are fully recyclable or
biodegradable, as well as a whole new paradigm for designing components
by adopting a "cradle-to-cradle" philosophy that supports the
remanufacturing of components from spent products into new products.
Evolving innovative technologies for effective and efficient recycling of
materials towards this objective is one of the foremost challenges.

Provide Access to Clean Water
About 1 out of every 6 people living today do not have adequate access to
water, and more than double that number lack basic sanitation, for which
water is needed. It's not that the world does not possess enough water; it is
just not always located, where it is needed. Though there is overall
equilibrium, effective & efficient distribution of water in order to eradicate
flooding as well as drought is the real challenge to engineering scientific
community.

Prevent Nuclear Terror
Engineering scientists share the formidable challenges of finding the
dangerous nuclear material in the world, keeping track of it, securing it and
detecting its diversion or transport for terrorist use.

Manage the Nitrogen Cycle
Human-induced changes in the global nitrogen cycle pose engineering
challenges just as critical as coping with the environmental consequences of
burning fossil fuels.

Develop Carbon Sequestration Methods
The growth in emissions of carbon dioxide, implicated as a prime
contributor to global warming, is a problem that can no longer be swept
under the rug. Cement industry is one such examples and evolving
technological solutions for partial or complete replacement of cement in
construction is need of the hour.

Summary

The challenges describes above are only indicative considering the facts as
on date. Many others are indistinct and many more surely lie beyond most
of our imaginations. I leave you with these three quotes that have
animated me:

"To the Optimist, the glass is half full; To the Pessimist, the glass is half empty; To the Engineer, the glass is twice as big as it needs to be."

"Engineering Scientists like to solve problems; if there are no problems handily available, they will create their own problems."

"Engineering Scientists turn dreams into reality."

Grand experiments and missions of exploration always need engineering expertise to design the tools, instruments and systems that make it possible to acquire new knowledge about the physical and biological worlds. I hope these examples above cause you to pursue careers in engineering science.

About the author:
Palani graduated with a degree in Civil Engineering from PSG College of Technology. He has a doctorate from the Indian Institute of Science. He works as a Senior Principal Scientist at CSIR – Structural Engineering Research Centre. He lives in Chennai with his family.

Question & Answer

Which leader do you admire the most?

Andy Grove was one of the two executives who made Intel the successful company it has become. As a corporate executive, he was one of the best the world has seen. His autobiography is very inspirational. He was a poor kid from Hungary who came to the United States with World War II as the backdrop, and went on to become one of the most successful leaders of our era.

His book *'Only the paranoid survive,'* is a classic that emphasizes the need to be always looking at the future to drive continued success. I also like his book *'High Output Management,'* and would recommend it to aspiring managers.

I have had the privilege of working at Intel and have observed Andy Grove in action. I have practiced a lot of the principles that he enunciates in my own career and can say that it has helped me become a much better professional.

Amar Babu
Electrical and Electronics Engineering
Mumbai, India

II. BRANCHING OUT

FINANCIAL SERVICES

An Engineer turns Banker

Tracing a path from India to Indonesia, the author shares his experiences on how he ended up branching out of engineering into banking. Given the turbulence in the global markets today he provides valuable insights into what is happening in this all important sector and what we can expect in the future.

Introduction

There is a common saying that goes, *"it runs in the family."* A doctor wants his child to be doctor, an army man wants his son to join the armed forces, and an actor wants his child to be an actor. And here I was, born to a Banker and was destined to be a Banker. The difference was that my dad never wanted me to be a Banker and used to remind me that if I ever join banking I would be wasting my career after having earned a degree in Engineering. Today he is proud of what I am.

Philip Kotler, the Marketing Guru, wrote the marketing bible in which he introduced the 4Ps - Product, Price, Place, and Promotion. These 4Ps are on the fingertips of every individual who wants to make a career in Marketing. I feel this is applicable not only to marketing, but to each one of us who are a Product or Commodity: the Packaging or Promotion is the Education we obtain which slots us in the Right Place (Job) and the Price we get paid for it. Your educational qualification is only a *VISA* into your industry of choice. Every calculated step that you take will determine your future. Your future is in your hands and its only you that can shape your destiny.

Beginning of the journey into Banking

Let me start with a quick personal story, before I give you a sense of the financial services sector.

My father worked for a Colonial British Bank, Grindlays Bank. The logo of the bank was an elephant standing on its two feet symbolizing *"Royalty, Strength and Power".* I saw him grow in the bank over his 30 years of service in a culture, which had elegance and class and thus started my dream to be a banker...

Life always presents you with choices and you need to decide what is good for you. All through my life, I have been presented with 2 choices - Post

10th standard to join the science or commerce stream, Post 12th to do Engineering or CA. I chose the former.

Post Graduation, I joined Tata Power, just to ensure that I lived up to my father's expectation of working as an engineer. After having worked for a year, had the choice to either pursue my MBA in India or do a Masters in Engineering in the US. I chose doing an MBA in India.

Post MBA, my father was keen that I join a leading engineering firm and productively use both my engineering and marketing degree. The dilemma continued... there were no banks coming to the campus for recruitment. I was offered a job at Godrej in their Refrigeration Marketing division, which was my second specialization in Engineering. I took up the job and joined Godrej and worked with them for three months.

Anyone in my place would have chosen a Commerce and a Chartered Accountancy to get into banking, but you can see that my dream of becoming a banker did not fade, even though I took a different educational route.

As destiny would have it, got a call from ANZ Grindlays Bank in late1989 asking me to join them, thanks to the summer internship I had done in their investment banking division during my MBA. I received a call from the General Manager of the Information Technology unit who wanted me to join this group. I readily accepted despite a bit of opposition from my dad who even refused to put in a word given his strong principles. But as destiny would have it even before I accepted the offer, there was an opening in the investment banking team and the General Manager under whom I worked for the project offered me a role in the investment banking team and thus began the banking journey in 1989.

As a management trainee, I had to go through a 3-month internship program with a group of 30 other people who were from IIM. We went through an intense training program, which were a mix of hard work and some fun. I still remember the Foreign Exchange training where we were woken up at 3 am in the morning with flash news to be used for taking / squaring our Foreign Exchange positions across our team mates. Our 3-month training was spread across Mumbai, Delhi and Chennai and staying at the Oberoi in each of these locations, which was a big luxury those days.

And then came the day when I was assigned to my unit – Capital Markets. I was thrown into the sea and had to start swimming from day 1 in the midst of the equity capital market sharks. Each day taught you something new and what helped me the most was quite a few of the staff had worked for my dad and had great respect for him. *The lesson I learnt at that time was it pays to be good to people as you move up the corporate ladder.*

Shocks to the system

I have seen the High and Low of the banking industry in the last 25 years. As they say there is a 4-5 year cycle for every product and the banking sector too has witnessed these cycles with many events that have affected it over the last 25 years: The Gulf War 1990-91, The Asian Financial Crisis in 1997, The Dot Com and Tech Bubble burst in 2000, 9/11 attack in 2001. The SARS outbreak in 2002, Global Financial Crisis in 2008 and now the ongoing European Debt Crisis with a potential Greek default.

Prior to every event the market was on a high and with each event we saw both the equity and debt markets going into a tailspin. The crash was so rapid that traders were like headless chickens reeling under heavy losses. Banks were forced to take prudent measures to cut loss positions. The fact that we went through so many low cycles made us more resilient to the shocks and with every event we would grow much stronger with an ability to face the worst. We would remain liquid and well capitalized to take advantage of the shocks.

The key was we used to treat each shock as an opportunity and post the 2008 global financial crisis we decided to enter aggressively and captured market share from our competitors. For a decade from 2003 – 2013 we grew at high double digits even during the global financial crisis.

Your personal and professional life is not always a bed of roses. You will face thorns or obstacles in each stage of your life and every obstacle you overcome teaches you a new skill, which makes you a stronger person who is willing to accept challenges. The same applies when you rise the corporate ladder, with every move up the ladder you will be faced with new challenges both on personal (interpersonal rivalry, jealousy, etc.) and professional (competition, strategy) and you need to overcome each of these and rise up to become successful.

Banking and Financial Markets

Over the last decade the banking and financial sector has seen unprecedented change that the industry is anything but the same. Banks have been trying to restructure themselves in the face of significant headwinds. The regulatory rulebook is being rewritten, the industry has suffered major reputational damage and major economies have struggled to emerge from recession. Further, the rising cost of doing business, regulatory requirements for maintaining additional capital and liquidity, changing consumer behavior (move to mobile and online channels) has resulted in banks taking a beating on ROE (return on equity). The banking industry is facing intense competition and is in a phase of consolidation and systemic stability. The key focus is on technological innovation, disintermediation, cost efficiencies and shareholder value.

Equity markets have hit record highs but on the other hand economic outlook is still not positive and questions are still being asked: *Is Europe coming out of recession, impact of the US debt ceiling, will "Abenomics" kick start the Japanese economy, will China's bubble make it go the way of Japan?*

Banks globally will need to incorporate enough flexibility in their models to respond to the new rules keeping in mind the future uncertainties.

The five forces, which will drive banks to change over the next few years, are:

Regulations: New global regulations will challenge the profitability of the banks.

Customer: Greater transparency, personalized products and seamless transition between channels to ensure solutions and a good product service mix.

Technology: Exponential growth of data collection, storage and analysis, data privacy and cyber security.

Competition: Battle for scale and market leadership, payments and lending solutions.

Society: Regulatory and Compliance culture will need a behavioral change, shareholder pressure to deliver sustainable returns and reward investors.

The global and local regulatory environment will necessitate banks to comply with requirements for capital, liquidity, leverage, recover and resolution planning. Banks will need to do a strategic assessment of their Business Model, Customer Relationships, Organization Structure and Infrastructure and effect a transformational change under the regulatory constraints:

With regular shocks hitting the market over the years, Regulators are ensuring that Banks become more resilient. As reforms kick in, banks have started to focus on their core functions and aspects of their value chain that have a competitive advantage thus restoring reputation and profitability and support economic growth. Banks also need to collaborate with third parties to deliver their new business models more effectively.

Are Banks needed?

In spite of the bad name big banks have been assigned due to the great crash of 2008, in my view, banking is still a necessity and with technological advances the demand for quality professionals continues to grow. While the heyday of banking perks (stock options, performance bonus) are history and remuneration is being capped by regulators, it still ranks among high on the pay scales across sectors.

Banking is also opening its doors to highly skilled and qualified professionals from other sectors such as technology, consulting and services. The days of the highflying investment bankers are slowly coming to an end and we are moving back to the traditional and prudent ways of commercial banking. As they say "old wine in a new bottle" still tastes good.

Summary

I have given here, an account of my personal journey in banking, even though I did not get the traditional education for a banker. I have also shared how my career progressed. Finally, I have also given you a sense for where banking is headed.
My mantra for success that I would like to share with you can be summarized as the 3Ds: **Destiny, Dream, and Deliver.**

Think of a Destiny - i.e. the goal or Aim in your life. Dream your destiny every day and ensure you *Deliver* to achieve your Dream and Destiny.

As our departed and legendary President Abdul Kalam said *"Dream is not that which you see while sleeping, it is something that does not let you sleep."*

About the author:
Prakash Subramaniam lives with his family in Jakarta, Indonesia. He works as a managing director at a global bank. He graduated with a degree in mechanical engineering.

Question & Answer

Which leader do you admire the most?

K. Kamaraj, is my favorite leader, hero and role model all combined into one. The more I read about him, I think he is the perfect example for a perfect leader for aspects that I would summarize in bullet form as:

Genuine, humble, visionary, selfless, focused, brave, sincere, hard-working, loving, people-caring, learning, recognizing, rewarding, team-working, guiding, helping, mentoring, problem-solving, supporting, forgiving, being-simple and communicating well.

Ganesh
Production Engineering
Singapore

INFORMATION TECHNOLOGY

Information Technology and Security

While the information technology wave put India on the global map after 'Y2K,' there were professionals finding their way well before that into this nascent field, which was about to explode. The author describes her experiences and insights into one of the most important aspects of this sector – security.

Introduction

When we were in college, Information Technology or Computer Engineering was not offered as a separate Undergraduate Engineering major at PSG College of Technology. I distinctly remember that the institution had just begun the Masters in Computer Applications program at that time, which had a good reception with applicants from around the country. In the undergraduate program, we had a few *electives* related to Information Technology, which students majoring in other areas could opt to take.

In the early nineties, Information Technology began evolving in many sectors in India, prompting many of us to switch over to it as a career. Of course post the so-called *Y2K*, the participation of Indians in this sector was a tsunami. I was part of the early wave of engineering graduates who began migrating to this field.

Many companies setup a separate Management Information Systems (MIS) Department. Software training centers mushroomed all over the country. The trend was to aggressively learn as many Application Software offerings as one could in a short time, and apply for jobs requiring those skills, preferably overseas.

The gold rush begins

Cramming resumes with a long list of Application Software names supposedly increased the confidence level in many candidates. Even the so-called software body shops encouraged candidates to match their resumes skill-by-skill to what was in the job requirement. *"So what if you don't know that exact skill? You know something similar right? Just put it down in the resume and you can learn the exact matching skill while on the job"* was their rhetoric.

What many did not realize, then, was that, it was *not* the number of Application Software types listed on the resume that would give candidates the strength needed either to secure the job or more importantly, do well once they start.

True value lies in understanding the fundamentals of the 3S's of IT, namely:

- Software Development Methodology (SDM)
- Source control
- Security

These *3S's* are platform and domain-agnostic, and are continually evolving.

An Information Technology Company whose primary business is software development may include the *3S's* as part of their training. But if we were to join the Information Technology division of a company whose main business is in some other domain, then the onus is on the applicant to know the significance of these fundamentals.

Hack attack

In 2001, I joined the Corporate Information Technology division of a USA based off-price retail chain. It was an $18 billion company then, with growing operations in the US, Canada, and Europe.

But on the Information Technology side, it left a lot to be desired. It took over 8 years, and a major hacking to put controls in place. It is still a work-in-progress as the company heads towards $40 billion in revenues.

Senior Management attitude before the hacking was *'we are not an Information Technology company, so it is not mandatory for us to achieve any high SEICMM/ISO standards. We would invest just enough in Information Technology to support our main focus, which is our retail business."*

The hacking was a painful lesson for the senior management in the company. After that episode, there was a radical change in the company's Information Technology policies.

A comparison of the Before-After scenarios would bring out the significance of each of the *3S's*, and how these add value to the overall performance of the company.

Impact of Software Development Methodology (SDM)

SDM is a framework for building and managing Information systems from inception to installation. Common methodologies include:

- Waterfall
- Iterative/Incremental
- Spiral
- Rapid Application Development and
- Agile.

Traditionally, many Information Technology companies used Waterfall to progress the Software sequentially through its life cycle, namely, Initiation, Requirements, Plan/ Analyze, Design, Build/Test, Quality Testing, UAT, Implementation and Post-Implementation support.

More recently, the Iterative and Agile methodologies are gaining prominence.

Our company had used Waterfall methodology, but not in its entirety before the hacking.

There were no well-defined teams and roles. Users would provide just the bare-bone requirements, which were often ambiguous. They would keep changing it numerous times, even in the later stages of the project, thereby leading to cost and time overruns.

Change Requests were not formalized. As a result, there were disconnects between the requirements and the end product. Information Technology managers would give into pressures from the business, and try to cut corners. Code may go live without thorough testing causing fatal defects/ failures in the Production environment.

Managing Source Code Control

Version control of Software is very vital. There was no designated Release management team before the hacking. Some developers did their code changes in a shared drive. They even managed to go-live with their code, bypassing Source control.

Some used Source control, but would have the code checked out on their machines for a prolonged period of time. Some left the company while code was still checked out on their machines. This resulted in Code overlays or situations when only the executable existed, but the Source code could *not* be found or had to be re-engineered.

Importance of protecting Security

Although our company managed Key Financial Applications in a more secure environment, still, as the importance of security was not stressed long and hard enough, it left the company exposed to vulnerabilities.

And we were not alone in being so exposed. Below is a list of major hacking episodes worldwide in recent times:

Heartland Payment Systems 2008: 134 million credit cards exposed through SQL injection to install spyware on Heartland's data systems.

TJX Companies Inc.2006: 94 million credit cards were exposed.

Google/other Silicon Valley companies: 2009, Stolen intellectual property

ESTSoft: 2011, the personal information of 35 million South Koreans was exposed after hackers breached the security of a popular software provider.

Target, Home Depot, Sony, 2014

And the list goes on.

Cost of a data breach - Studies have shown that the average cost of a data breach is around $7 million with average cost per compromised record more than $200.

The remedy

The after-scenarios in our company showing improvements in all the *3S's* I have described earlier.

- Firewalls have been strengthened
- People who exchanged Credentials met with severe penalties
- Credentials were reset when job roles were changed, and removed when people left the company
- Mandatory Security training for all Information Technology associates conducted periodically
- Bit-9 Parity checks to stop anyone from installing unauthorized software on their machines.
- All across IT, well-defined teams formed with clear roles and responsibilities.

- RACI matrix looked up to see who is **R**esponsible, **A**ccountable, **C**onsulted or **I**nformed for a specific task.
- Release Management team put in place checks and balances for any software that is released to the live environment.
- Release Management published the guidelines for parallel development, and branching and merging code in central repositories.
- Estimation is done at various stages of the project, and budget increases are strictly scrutinized.
- Any lapses in the process, escalated to senior management. Also publicized widely to name and shame individuals who are repeat offenders.
- Projects for total encryption/ masking of cardholder data undertaken.

For individuals the future is..... a tad scary!

It is important to emphasize that this sort of vulnerabilities are not just for companies but or individuals also.

Marc Goodman is the founder of the Future Crimes Institute and a former INTERPOL and FBI agent. He lists a litany of threats facing us. Here are some examples he shares of what has already happened in terms of crimes:

Automated Teller Machine (ATM) Thieves: Fake ATMs where installed at a shopping mall. When shoppers used these ATMs, their card numbers and pin's were recorded. They got an 'out of order' message on the ATM. The thieves then used these numbers to withdraw cash elsewhere.

App Thieves: Recently there was a fake HSBC Bank app on Android. Similar to the above example, people entered their data into it and were subsequently robbed.

Telephone Network: The Mexican drug cartel has its own installed telephone network and use auto piloted vessels to transport drugs into the US

It is worthwhile noting, that while individuals may not be able to take as stringent measures as companies can, it is imperative that we take a commonsense set of precautions as well such as being careful with passwords, protecting private information carefully and not leaving private information lying around.

Summary

I have pursued a career in in Information Technology and through its course recognized the importance of sound security measures. In today's global economy, individuals are interconnected, companies are interconnected and economies too are interconnected. This is bringing enormous innovation, scale and access across the globe.

However, these very same advantages are showcasing some vulnerabilities as well. As I have explained in my own company's example above, those businesses that do not have proper security measures, that evolve with the times, leave themselves vulnerable to attacks. Not only are companies vulnerable, so are individuals.

Within the secure domain of companies, having a holistic perspective of the *3S's* (Software Development Methodology, Source control and Security) along with strong technical and domain knowledge is paramount. Such a focus for those interested in Information Technology will pave the way for a sound career in a vital and growing area for enterprises.

About the author:
Bharani Rangabashyam lives with her husband and son in Leominster, Central Mass, USA. She works in Corporate IT for a multinational Retail chain. At PSG College of Technology, she was a day-scholar graduating in Electronics and Communication Engineering.

Question & Answer

Which leader do you admire the most?

I like the combination of faith-in-oneself like Mahatma Gandhi, the persistence of Edison, fearlessness of Indira Gandhi and following the inspirational lines of Swami Vivekananda.

Krishnan
Production Engineering
Bangalore

ENTREPRENEURSHIP

The Entrepreneurial Journey

This chapter demonstrates why entrepreneurship is a thrilling, taxing and rewarding path all at once. The author describes his journey from India to the United States and shares how he never lost his passion for starting his own company. Get ready for an exciting read!

Introduction

During my school days in Rajapalayam, several of my father's friends and my close friends were all from an entrepreneurial family and ever since those days, I always wanted to own my own "factory".

When I graduated with my Masters and on my way to get my MBA, I had an opportunity to present a paper in Statistical Process Control (SPC). In the audience, was the VP for Monroe Shock absorbers a division of a large conglomerate Tenneco. He approached me after the talk and wanted me to come and address his staff. The meeting he wanted me to come to was in a very nice resort in Florida and included golf. He even said he would pay me for the day. I never thought of my student visa and all the potential issues this might cause if I accepted a fee from a corporation.

My naivety and total eagerness to golf in Sawgrass, made me accept his offer. I scrapped my entire life savings of under $500 and bought an air ticket and landed in Florida. I don't remember how my talk went, though I am sure it was a total disaster. But on the golf course, I was in the foursome with this Vice President and he asked me to come in as a consultant for one year to help them implement their SPC program. I still did not tell him about my visa situation.

I started to work the following week and their accounting department asked me what my company was called so they can issue my check. Just like in the movies, I came up with the name. Though several months later, I got a notice from the Immigration department and had to tackle this, I learnt a valuable lesson here. *"When an opportunity knocks, accept this and then figure out later how to piece the puzzle together"*.

Starting a business

A few years later, the bug to start my own factory started to haunt me. All my friends kept telling me to accept the permanent job that was offered by another large automotive company. It was summer of 1993 and my mother was visiting me. She knew me well and she said, *"Do what you want. What is the worst thing that can happen? You will lose all this money you have made, but you have a great education and I am sure you will get a job later on as well."* Mothers know us best.

I came to Augusta, Georgia to start *Palmetto Industries*. I had seen this product a year earlier when I was visiting a friend, with an ulterior motive to get a Sweat shirt from the Masters Golf Tournament, one of the four 'Majors' - the other three being the US Open, the British Open and the PGA. Though I returned empty handed back to Michigan, I knew I could make this *Flexible Intermediate Bulk Container*.

So, in 1994, all my belongings and myself in my two door car made it to Augusta. Again, naivety and just my passion took over all logic. After 20+ years, I have been asked my classmates to write this chapter. I must have done something right. Truly, this opportunity means a lot more than several of the accolades I was given.

Golden Rules of business – Will you break them or not?

During my first few weeks I was in Augusta, I met the just retired President of University of Georgia. Dr. Fred Davison. He was larger than life and I could never get to call him Fred, as he insisted several times. So we settled and I called him Dr. Fred. He told me the following rules. Though I never wrote them down, I remember them and I have attempted to capture his advice. He also told me I had to follow this at any cost.

- Hire the best employee. Interview a lot, check references, take your time but once you hire treat them fair and commit to a long relationship.

- Ensure your employees are smarter than you and listen to them, but make your decision based on your gut and their intelligence

- Treat your suppliers better than your customers – Customers will leave you but suppliers will want your business and will work with you

- Never compromise on your quality – Never

- Take your time and come up with you corporate values and vision but never compromise

- Hire a good lawyer and an excellent CPA firm. Also make them as part of your board or your advisory board.

- Never underestimate your competitor and never get a business just because you want to take it away from a competitor.

- Innovate, Innovate and Innovate – Don't offer a "me too" product

- With every paycheck period, put a small amount into a savings account. Start with $100 and make it higher as your business takes off.

- Learn all aspects of your business – Nothing in your company is too small for you to do or too big that you cannot do.

- Commit to your community and always participate in local area politics

And then he said…

- Nothing will ever work as you planned.

- On this last point the boxer Mike Tyson has a famous and funny saying, *"Everyone has a plan 'till they get punched in the mouth."*

Did I always follow the above rules? I have tried to and may his soul rest in peace. If I can dedicate this chapter to Dr. Fred and have him send me his blessings, I would be grateful.

Lessons Learned

Hiring and layoffs
One thing I learned was though I always had the best intention while hiring someone, if it does not work out, let them go quickly and gracefully. Dragging this out is not good for either party.

Listening
Listen to your employees. BUT verify several times. If necessary bring in outside consultants to validate your decisions.

Suppliers
Suppliers are lifesavers - But always negotiate the best value for your company. During the 2009 financial meltdown in the USA, I had to go to several of my suppliers and work out a payment plan for past dues while ensuring my supply was going to be uninterrupted. Being honest 100% was the only way I was able to keep my company from bankruptcy.

Quality
In our space as in several others, quality is a subjective term. I used to supply some Poly Ethylene Tarps to a very large big box retailer in 1996. I used to import this from South Korea those days. The specifications from the customer, which the supplier agreed to was far different from the actual product. My customer and my supplier and I all knew this, but nobody said anything, because it was a commodity item.

After two years of the specification continuing to degrade, I had to confront my customer and ask how to tackle this. I was shocked when he said, "*I always knew what I was getting, but the label says what I want it to say - So don't worry, nothing will happen to you.*" I had to listen to my team and my conscience and next year did not bid on his contract.

Thankfully, some changes in his corporate program, they revised their terms and specifications to match the supply and five years later asked us to bid again. This leads to a very good question – How much do you educate your customer? Do you want to be 100% transparent or only tell them just enough? Over the course, I have adopted the latter.

Flexibility
As a founder and entrepreneur, always be flexible. In 2008, we launched a new product. It was our idea to take this to the market all by ourselves and did not want to go thru distributors. I even brought in a new Director of sales. After spending over 12 months in marketing and meeting with several customers, we had very little traction. My entire team was still very optimistic and we knew we could break the barrier and go at it by ourselves.

The 2009-year started out extremely bad and I was worried. In the meantime, I was approached by a large distributor and a competitor. Over a few drinks and a round of golf, I had signed an exclusive agreement. I had to call my new Sales Director the next day and tell him. I had underestimated the resources required to penetrate this new market and had to change course. I had to do what was good for everyone. Sometimes, you make a great team even with your competitor and is a WIN-WIN for all.

Communication

I always start my meetings by asking all to first tell me the bad news. Good news always travels fast and everyone wants to brag about this. Prior to me starting my company, I was doing some consulting work for a bank. The CFO of this bank always said "Air your dirty laundry, don't sweep it under". It is very critical however to not point fingers or to lose focus of the meeting. I view this like talking to your kids. They should be afraid that Dad will punish them for something they did wrong but NOT be afraid for telling me what they did. The punishment for lies or "sweeping it under" is worse. Also once the issue has been laid out it is everyone's issue and a team approach is needed to solve this. I also never hold these against the managers during review.

Emotions

Never do take any decision when you are angry, happy or emotional. Invariably you will make the bad choice.

Work Life Balance

Always put family first over work. I always take my wife's call and always set make my travel based on my kids schedule. Nothing is more important to me. I found out when I told my customers why I was not able to take their call or visit them, I actually scored points. How about that? Honesty actually paid off!

Technology

Embrace new technology. In 2000 or so, one of my colleagues on her own initiative developed our own ERP system. It was written in MS Access. We quickly out grew this. Around this time, the Internet was taking shape. One our classmates – Sudha came to my rescue. Very early on, she recommended I make this whole thing web based and not resident software. Take full advantage of social media for marketing. Equip your sales staff with all the fancy gadgets for presentation

Is it Fame or Fortune that drives an Entrepreneur?

I think neither and I feel it should be neither as well. Though both will seek him for a job well done, what drives him is the passion. I remember the golden rule – Innovate, Innovate and innovate – This applies to all areas of the company – from product innovation, to process innovation to customer service to logistics. I keep tweaking the company constantly and sometimes drive my colleagues insane. I try to stay a few steps ahead of my competition and I feel it is my job to do so. Of course a nice pay check at the end of the year helps and especially as I get older and retirement staring at me, fortune starts to play a bigger role.

One of my corporate values is I will never "bribe" a customer or "buy" some business. In 1999, I found out one of my sales guys was having illicit dealings with one of our customers. The account at that time was almost 20% of our gross sales. Nevertheless, I had to fire the account manager. But I did not know how to deal with the customer and had to write to their President. I lost that account and have never been able to get it back. My account manager went to work for my competitor and guess who has that account to this day?

Exit Strategies
Every entrepreneur needs to plan his exit – Does he hand over the reins to his kids or go public or sell. I always planned to sell and hopefully if we have another book release in 5 years or so, I might write a chapter in a successful merger!!

Role Model
Everyone needs a role model. I always liked Steve Jobs – His passion to build the best product and never compromising on quality was what I started to admire him for. But his second act, as he took over the reins back from the company he founded and made it a truly world class was made me almost worship him. In 1990 or so as a graduate student with just a few $100 in my pocket, I made my first stock purchase – it was Apple stock. I think I bought 50 shares or so.

Am I glad I kept it to this day!

Summary

I have written this chapter based on my experience with a hope there are some gems for the future graduates. Whether you are going to be an Entrepreneur or go to work in a large company, I feel the challenges are the same. Several of the above rules apply to all. We should do what excites us and keeps us challenged. When the day comes we are working for the next paycheck, that is the day we need to take stock and regroup.

About the author:
Shankar Balan is a Mechanical engineering graduate. He currently lives in Augusta, GA USA with his wife and twin children. His work is in the packaging industry.

Question & Answer

Is employment better or entrepreneurship?

After all these years of working I am tending towards entrepreneurship because of the sheer joy of creating something that you can call your own despite knowing that I may have sleepless nights!

Mohan Alapat
Mechanical Engineering
Bangalore

ENTREPRENEURSHIP

The Customer is the Real Boss

One of the most important questions in business is its purpose. If one cannot answer the question, 'Why are you in business?' everything else will be very hard work. The author, who is an award-winning entrepreneur, describes what motivates her.

Introduction

I graduated with a degree in Electrical and Electronics Engineering from PSG College of Technology. I am an entrepreneur in the supply chain and distribution business based out of Thanjavur. My business is called LPG Distribution System and we distribute cylinders. I am proud on behalf of my company and my hard working associates to say that we have received many awards for customer excellence.

I wish to share our lessons learned on customer excellence, here with you. I strongly believe that whatever business you may be in, by definition, you will be dealing with customers, and thus if you make customer experience and customer excellence a core practice, you will have a key element of success in your arsenal.

Importance of Customer Experience

Sam Walton, the legendary founder of the Walmart superstores, once said: "There is only one boss. The customer. And he can fire everybody in the company from the chairman on down, simply by spending his money somewhere else."

To use an analogy, one can imagine a business to be akin to an online portal. Customers are in the front end, company employees are in the middle tier and the employer is at the back end. We all know from our own experience that if the web pages we visit are not user friendly, impressive, informative, current and attractive, we will not visit the website again.

Similar to this, employees in an organization should be willing to serve with a smile, provide the customer a great experience, should have the thorough knowledge about the product they are selling or servicing and should be empowered with the ability to supply the customer's needs and wants.

It is not enough to 'check the box' in terms of a customer service process. If the customer is not satisfied they will never come back. The core of good customer service is providing the customers more than what they expect and to give them a *wow experience.*

Only excellent customer service will help the organization to retain the existing customers and to get new customers.

Building a foundation for world class customer experience

There are several key elements for building a foundation for world-class customer experience including focus on recruiting and training that are explained here.

Hiring for the Attitude

When hiring new employees, we need to be sure to select the one who is willing to go the extra mile over someone with greater technical skills. Attitude is important than aptitude. I have found in my experience that it is much easier to increase the technical skills of an employee but is difficult to change their attitude.

Employees who excel at customer care have a natural desire to serve and express genuine empathy when conversing with upset customers.

Once a young female candidate approached me and requested me to give her a job. She was a school dropout and she told me that she doesn't know anything about computer. I asked her whether she is willing to learn, she said yes. I appointed her and within few months she learnt the entire process. She is one of the very successful employees at my company.

During our school years we had a few courses in business management in addition to our courses in engineering. When we were younger, we focused on the engineering side of our studies and paid cursory attention to management insights. Now, after all these years, I recollect and value what we were taught then about employee turnover and what various best practices we need to put into place to ensure that we retain our best employees.

The opposite of this is also true. Netflix is a popular online video streaming service in the west. I highly recommend reading a publicly available presentation by the CEO of Netflix, Reed Hastings. He talks about how we proactively transition employees who do not fit the innovation and customer service mindset at Netflix.

Thus both insights are valuable - a person with positive attitude is an asset for our company. Treat them well and retain them. And a person who does not fit the culture of your company, be gracious, but help them find a different opportunity.

Training the Employee

George Patton, the great World War II general, famously said: *"The more you sweat in peace, the less you bleed in war."*

Employees, who are properly trained will have professional customer service skills, they can satisfy the customer and gain their loyalty. This helps the business retain customers and improve profits. It costs less to retain loyal customers than to acquire new ones. Satisfied customers will trust us and they will refer about our company to their neighbors, friends and relatives. Investing on Employees training will only fetch fortune for our company.

Here, I would like to introduce to you the concept of *Net Promoter Score*, created by the management consultant, Bain & Company. If you ask you customers only one question, it would be this. Will you recommend our business to your friends and colleagues? The net promoter score is the difference between those who say yes and those who say no. Business research has proved that this is the best predictor of how well your customer service is doing.

In some organization the policies of the management will tie the hands of the employee. Therefore it is equally important that management also needs training. Employee empowerment allows staff members to make decisions without consulting their bosses or managers. By being allowed to make choices and participate on a more responsible level, employees become emotionally bonded company. They often view themselves as representatives of the company. This helps in redress the customer complaints fast.

A great example of this empowerment is the Ritz Carlton hotel chain. At the Ritz Carlton every employee at the hotel from the manager to the janitor are empowered to make decisions unto a certain monetary amount. If they are faced with the customer, and they need to address a customer complaint, they are empowered to assist the customer, without checking in with their supervisory chain of command.

Closer to home, in our own organization, I very much take the attitude that

I will not isolate myself and sit inside the cabin. I am constantly at the front desk and share the work with my staff.

I try and not interfere when they are handling the customers. Only when it goes beyond their limit to handle the customer I will take assist them. In some situation our staff will handle the situation better than us. They have trained me several times.

We mutually teach each other

The management consultant Phil Kight teaches what is called *'above the line'* behavior. That is, given any situation, we all have a choice - we can either behave above the line or below the line. Below the line is destructive behavior. Above the line is constructive behavior. Above the line builds a better culture.

Using this as an example, I aspire to be a role model for my staff. More than what they learn from training they tend to learn from observing how others behave. If their managers are a disorganized mess, so will the staff. If others have a negative attitude, so do the rest. If we hide the truth and cut corners, so will the staff.

On the other hand, if we can be calm, confident, hardworking, meticulous, and fair and a problem-solver, then each and everyone of the staff will emulate that behavior.

Gaining total customer satisfaction

We know that satisfied customers will buy more of our stuff thus generating higher profits. How are we going to satisfy the customers? First we must identify the need of the customer. In our LPG distribution system, timely delivery of cylinders with quality and quantity is the prime requirement of the customers.

Our brand promise is that the product will be delivered within 48 hours.

The key factors that are involved in executing the Brand promise include:

1. Inventory Management

2. Delivery system including Delivery Men and Mechanized Vehicles.

3. Climatic Conditions.

Only when the key factors, which play major roles in our organization, work together it can result in customer satisfaction.

Keep measuring the performance. Give incentives to the delivery staff that help to improve the measure.

In our business customer satisfaction is not possible always. We are distributing subsidized product, which is meant only for domestic purpose.

Some of the customers who try to divert that for commercial purpose will hit the landmine of dissatisfaction frequently. In a different chapter, my classmate Subramaniam will be discussing ethics in business practice. I highly encourage you to read that.

On the subject of business ethics, we cannot remain silent thinking that the customer is always right. We have to be tough with those customers if they are not willing to correct themselves, knowing well that they are going to spread negative publicity about the company. To overcome this negative impact, our customer service has to be exceptionally good.

Customer Loyalty

Satisfied Customers are like a rear view mirror who will give the picture about the past. Only the loyal customers will predict the future of our business. Satisfied customers will migrate to your competitor if they are going to give a better service than what you give.

In our organization we used to conduct lots of customer loyalty programs. We used to go to colonies and will organize free service camps. During festivals we used to join them and celebrate it. We used to organize lots of game shows and make them enjoy the day with us. As a part of our social responsibility we organize medical camps, educate the customers about safety and conservation.

Such initiatives help us a lot to interact personally with most of our customers and gain their loyalty. Loyal customers are our brand ambassadors.

Customer Complaints

Customer complaints are free survey that helps us to identify where we need to focus and improve.

For every customer complaint, there are six other customers who have remained silent. If we try to solve the problem pertaining to complaining customer it is going to solve similar problem faced by many other, silent customers.

Customer complaints help us to learn what we lack.

If you do not listen to the customer complaint and solve it, the customer will take their business elsewhere.

If you put utmost effort to solve the issue it will increase the credibility of the company.

Summary

As you can see, I not only value my degree and learning from college, but I am quite passionate about my business and my colleagues. But above all, I am passionate about customer excellence. Over the years, on the road to being recognized for this area of my business, we have learned many lessons. I have tried to distill those lessons and share them with you here. Please remember, the customer is the real boss!

About the author:
Sumathi Singaravadivel is an entrepreneur and head of a distribution agency called LPG Distribution System based in Thanjavur. She graduated from PSG College of Technology with a degree in Electrical and Electronics Engineering.

Question & Answer

Is employment better or entrepreneurship?

Being an entrepreneur is like having a disease. You pick up this bug during your early school years and then there is no cure. Many great companies that exist today were not started with a vision but with someone's dream to make a product better. This dream consumes the founder and all problems and issues that he/she faces are pocketed and fragmented in his mind and the only goal is making that product.

As long as I know I am building that better product, I am satisfied and feel today was a great day and tomorrow will be even better – I know people around me will see what I saw yesterday, today and their tomorrow will be better as my today was.

Shankar Balan
*Mech*anical Engineering
Augusta, USA

III. REFLECTIONS

LEARNING

Differentiating between education and learning

The world is changing rapidly and along with it technologies, companies and entire sectors. What does a professional need to navigate this in this world? Is it enough to be satisfied with a degree and go out into the wide world? Or is there another core skill that is needed? The author delves into this key question.

Introduction

Ever wondered why people ask, *"How much are you educated?"* No one seems to ask, *"How much have you learned?"*

I wonder if it is because they only want to know about the formal education that one has got, and not about the knowledge that one has gained over a period of time. Or perhaps it is easier to make assumptions about a person based on the degree they have attained or the 'brand of institution' they have attended. Whereas it is more complex to judge how much learning one has - for the questioner has to be equally competent to judge the extent of the learning.

Something and Everything

One day in 1983, when entering the hostel/dorm at PSG College of Technology, I saw on the black board a statement *"An Engineer is one who knows Something of Everything, and Everything of Something"*.

This was written by Prof. Venkatraman, then Principal of PSG College of Technology.

This statement looked like a statement of word play in English, but it was thought provoking. I internalized it, mostly because it sounded nice. Looking back, I consider it a cornerstone in my building.

It helped me in clearing the GATE entrance examination for getting into the Indian Institute of Technology (IIT) for Masters degree.

I used to watch the TV program in *Doordarshan* named the "UGC program". This was an educational program by the *University Grants Commission*. That time there was only one channel, and we did not have the luxury of flipping channels, so I had to watch this program while having the evening tiffin. This UGC program used to cover educational programs in all

87

kinds of subjects and streams varying from engineering, science, medicine, mathematics, sports and geology. Watching the audiovisual programs let me learn these programs easily without me having to effort consciously. I did not think that at that time, it would help me someway in the near future.

Entering the GATE

The time came soon. In the final year, I wanted to attempt the GATE exam and get into IIT for higher studies. I was determined about this, and I had prepared well, even to the extent of neglecting to study hard for the final semester papers. The GATE exam had 2 papers, one paper in the morning, wherein the student would have to attempt objective type questions in 4 different subjects including mathematics, physics, thermodynamics and structures and one paper in the afternoon on the main branch of engineering that one studied e.g., Mechanical, Production, Electronics and so on.

For the morning paper, I had prepared for 4 subjects, Physics, Structures, Mathematics and Thermodynamics. After answering the questions in Physics, and Structures reasonably well, I was stumped with the questions in Mathematics and Thermodynamics. I could not understand anything other than the question serial numbers. What can I do, now that I cannot answer 2 subject questions that amount to 50% of the total marks in paper 1?

"*Stay calm*", I told myself.

Luckily then, while we filled the application form for GATE, we did not have to pre-select the subjects that we would attempt. We could select and write, whichever subject we wanted during the exam time. I remembered this, and scanned through the different subject questions. I realized that I could recall the answers for almost all the questions in Metallurgy though I had studied this subject in the 3rd semester. That done I again scanned and found that I could not answer the subjects like chemistry that I had studied earlier.

Was my luck running out?

Running out of options, I saw the subject Geology. Having nothing to lose, I scanned the questions in geology. I could visualize the UGC grant commission program popping in the eye of my mind giving the answers to the questions. *Voila!* I could answer geology questions, and there I wrote the answers to the geology questions.

I scored well in GATE and went on to do my masters in Industrial Management in IIT Mumbai, then known as Bombay.

What I learned "something in geology" then, helped me then in GATE. I owe it to my internalization of the statement *"An engineer is one who knows Something of Everything, and Everything of Something"*.

Learning by doing

When one reads the learning is limited. When one observes how something is done the learning is better. When one learns by doing practically the learning is complete.

This reminds me of what once Sri Ramakrishna Paramahamsa said in reply to a question as recorded in his Gospel, *"What is the use of describing milk - that it is liquid, that it is white, that it tastes nice or that it is good for you? The only way you get value from the milk is drinking it"*

At PSG College of Technology, we had this unique opportunity to learn by observation and learn by doing practically. The college had on its campus the PSG & Sons Charities Industrial Institute. During our engineering study, every week we had a 3-hour visit to the PSG & Sons Charities Industrial Institute. This was a manufacturing factory producing pumps, motors, agricultural diesel pump-sets, Lathes, Drilling machines and other kinds of machine tools. This industrial institute had a foundry, pattern shop, machine shop, motor winding shop, lathe assembly section, heat treatment section, Computer Numerical Control (CNC) machines, and robots.

What better opportunity could an engineering student ask for?

Each week we were assigned to one shop to observe and learn. Sometimes we were given small assignments or tasks such as:

- wind the stator & rotor coils for motors
- sit at a stamping machine and produce the motor stator or rotor stampings
- assemble the headstock of a lathe
- make a sand pattern in a pattern-making machine
- make an oil core for a molding, count inventory.

When I close my eyes, I can still visualize the shop floor and the work that we did. Did we learn?

I guess we learned more by observation and practical work that has stood as our foundation.

Pareto's Law

With that foundation, later in my first job, I could stand up to the challenge thrown by the shop floor workers and prove that I am not just a *text-book engineer*. I could handle a machine and run it like a professional machinist or could spot a mistake in a CNC program and correct it quickly before it ruined the job.

- The purpose of Reading, is to Understand.
- The purpose of Understanding, is to Absorb.
- The purpose of Absorbing, is to Internalize.
- The purpose of Internalization is to Practice.
- The purpose of practicing is to realize its benefits.

My professor of Materials Management at the Indian Institute of Technology, Mumbai taught us with the help of case studies. He used to give us case studies that we had to solve and present in the class. At the end of our presentation, he would summarize the key aspects of the chapter in question. Due to this case study method of learning, the lesson was retained well in our minds.

I got the opportunity to take it a step further, to practice, to implement in real life and realize its benefits. While I was learning materials management from my professor, I was the hostel mess coordinator in my hostel-2 at IIT Bombay.

In the hostel cafeteria, I introduced the concept of 2-bin system of ordering. For example, the monthly stock of rice would be kept in 2 different locations. In the first location would be kept 3 weeks consumption of rice, and the second location would be kept 1-week consumption of rice. The issue of rice initially would be from the 1st location.

When all the sacks of rice are consumed from the 1st location, the reorder would be triggered and the hostel manager would place an order for the next 1-month stock of rice from the merchant. By the time the stock arrived, the rice from the 2nd location would be used up almost. This way we could ensure a timely ordering of rice without having to wait till the last moment to order rice at a premium price.

One of the initiatives was to reduce the monthly bill incurred at the cafeteria. How could we achieve it?

Again I put into practice what I learnt in materials management lessons. *As per Pareto's law, 80% of the cost is due to 20% of the materials.*

Adopting this, I found out that the most significant cost is due to cooking oil consumption. We changed the menu appropriately to use up less oil, and over a period of 2 months reduced the cooking oil consumption by 50% and this reduced the monthly cafeteria bill.

Lifelong Learning

Learning is a continuous activity, and is a never-ending one. A best practice is to keep a regular 30-minute learning period every day and stick to it. One could read any subject matter in that 30 minutes and grasp whatever is possible to be grasped in that time period. This would help in "Knowing something of everything."

In history there are many famous personalities who did not have a great education beyond basic schooling. If we look at these personalities to see how they gained success, it would be because of their ability to learn.

Education has its limits, learning has none

Education has its limits because there are only so many courses being offered in institutions. Education is that which is taught. Learning is by self. Hence learning has no limits of time, no limits of age and no limits of volume to learn. It is left to each of us to spend time to learn.

Unlike in the olden days of only frequenting the libraries, today the Internet offers limitless scope for learning. It can be done at any time and place convenient to each of us. Audio files, YouTube, Ted talks, blogs, Massive Open Online Courses (MOOCs) - the scope is large.

It is worthwhile remembering that we need to not only learn, but also internalize, absorb and practice and thus realize its benefits.

The need for a questioning mind

Once, my professor 'let me hear it,' for reproducing in my assignment paper, something from a journal verbatim, and even quoting the journal. He asked me did I believe in what was stated. I said I believed it because it was published in a journal of repute.

He then said that in the years to come there would be many authenticated and unauthenticated statements that one would come across. Unless we question them and weigh them logically and accept it through careful logical analysis, it would be dangerous for us to accept what is written in different sources. That was in the year 1987, when Internet and worldwide web was unheard of.

How true it is today in the days of unlimited search tools that churn up so much unauthenticated information!

What a thoughtful statement from him - *"Have a questioning mind."*

The questioning mind is one of the tools that I use regularly to get different answers and solutions.

Open book examination experience at IIT Mumbai

In 1987, in our Marketing Management paper, our professor gave us an Open book examination. He gave 2 questions to us, and gave us 3 days in which to answer. He told us that we could refer to any book or journal in the IIT Bombay library. It was the time before the advent of Internet, and hence we had limitations in our search facility. We had to resort to use our minds to search through different books and journals in the library to write the paper.

We all fared poorly in this exam.

What was the reason?

We did not know where to tap for the right information, and hence could not get to search in the right books and journals. Even today, if we do not have enough basic information on a subject we would end up getting ourselves wound up in endless searches in the Internet without much of success.

Our lesson: *Know where to tap for information, and what to tap.*

Prof. K. Venkatraman, who I have mentioned at the beginning, wrote a paper on *Theory of Machines*. Did he prepare us for the future?

In the third semester of engineering we had the paper he wrote on the

Theory of Machines and it was taught by the author himself. The subject matter was tough, but the classes were interesting.

When we opened the question paper in the semester exam, we were shocked. The questions were not the simple straightforward questions from the classic textbooks that we had gone through. Each of the questions incited us to think about different concepts that we had learnt in Theory of Machines and come up with the answer.

For example, one question was about the testing of a diesel engine on an engine test bed. Usually this kind of question would involve the calculation of Brake Horse Power of the diesel engine.

This question was asking "If the brake drums are cooled by water at a certain flow rate, what is the final temperature of the water running out?" This involved us to go outside of Theory of Machines and use the concept of Heat Transfer and Fluid flow rates along with Theory of machines.

I wonder if he challenged us so that we prepared ourselves for the future by knowing how to connect the dots.

Let me share a touching anecdote here: We had gathered in 2011 for our 25th anniversary celebrations on campus. A small group from the class of 1986 decided to visit Prof. K. Venkataraman at his house. He was retired and quite elderly by then. He had tears in his eyes when students visited him out of sheer affection. Many stories, laughs and memories were exchanged during the visit. This was particularly possible because he was also principal of the college at the time, and thus got to know many of the students well as part of these responsibilities.

As the group was leaving, ever the professor, he whipped out a piece of paper and gave the visiting group a design problem!

The rest of us who had not gone to visit the professor gave an affectionate but audible sigh of relief.

Imagination and Intelligence

"Imagination is more important than Intelligence," Albert Einstein.

How true. When the human being started to dream, started to imagine, new things came into being. New methods, and processes, new literature new discoveries and inventions came into being. Intelligence helped achieve

93

what the mind imagined. Without imagination there would not have been the possibility of creation.

So my encouragement to you is to start imagining. Get on the road to discovery and invention by starting to imagine. Only in our imagination can we set ourselves free of limitations. Once we get limitless thinking, the possibilities multiply, the avenues open up.

There are many intelligent people around in the world, and many of them do not get to achieve much. Yet there are some achievers who have not been considered particularly intelligent. These people had one great prowess, that of imagination. Once armed with imagination, they imagined what they wanted to achieve, and it drove them to achieve it by getting whatever skills that were required to master the achievement.

One such example is Wright Brothers who were only bi-cycle shop owners. They did not have formal learning in technology beyond bi-cycle repairs. Yet they achieved the first flight. If one were to look at the parts that made up their first flight, it consisted of bi-cycle parts. Their imagination drove them to achieve the first powered flight by human.

Imagination opens up a large space of possibilities.

Leonardo Da Vinci was a person with great imagination. Coupled with learning, he excelled in everything. Where he thought that he had to gain knowledge, he went and sought it.

Enough of just reading. Let us start thinking, start Imagining and start the action.

There would come a time when in some of the subject areas the reading material would become repetitive. It is then time to start to put into perspective all that we have read in that subject, connect the dots, and form our own thoughts and perspectives.

That is when original ideas spring up. Coupled with imagination it would open up a new arena to conquer.

From musical notes to music

I would like to share a particular aspect from the Southern Indian tradition of *Carnatic Music*.

Students learn from a teacher step by step. After a certain stage they are encouraged to experiment on their own. This is when small sections of compositions are made by students from their minds, from their imagination. This makes the rendering by each student different and brings a beautiful melody to life..

Similarly, each of us by the time we are in our twenties would have a good basic knowledge of some of the subjects. It is time to use up this basic knowledge and churn it with formulations and hypothesis to come up with new findings.

Examples are plenty:

- Coming up with a new machine
- A new mobile App
- A new management hypothesis
- Anew way of improving production performance
- A new art work
- A new architecture.

Once the thoughts spring up, it is time for Action. The action to achieve in real what we thought up is the culmination of our learning.

Summary

Throughout my schooling and subsequent career, from my first day at college, I have been struck by the importance of the difference between education and learning. I hope through many of the ideas, examples and anecdotes above, I have been able to share the crucial distinction between the two. One of the great writers of science fiction, Isaac Asimov said, *"Self-education, I firmly believe, the only kind of education there is."* So let me summarize by stating that education is something that is taught, and learning is something that is self-taught.

About the author:
Krishnan is a Production Engineering graduate. He currently is engaged with companies in India in improving manufacturing efficiencies through digital technology. He lives in Bangalore with his family.

Question & Answer

Is Employment better or Entrepreneurship?

Almost all entrepreneurs in my circle tell me something common. That there is a sense of gratification of contributing to so many families! That is unique I think. It cannot be so direct in a corporate life.

Venkat Raghavan
Electronics and Communication Engineering
Dubai

SPECIALIZATION

Which is the better choice - being a Generalist or a Specialist?

In everyone's career there comes a time to make a crucial choice. There is not judgment on the choice. But once made, it is hard to reverse. The author who has spent time at a major multi-national corporation deliberates on this choice.

Introduction

What do I choose to become? Which one would offer me more opportunity? Which one would pay me more? Which one will help me grow faster?

When you start your first job you may not ask all these questions as you may look for an association with a well-regarded brand. You might be focused on a particular industry or perhaps even a location.

However sooner or later you will be faced with the question of whether you become a generalist in your career or become a specialist. To understand these choices and the implications we need to start from the definition.

My early understanding of Generalists & Specialists

When I graduated as an engineer in 1986 from PSG College of Technology, I did not know what a generalist meant as I had not heard that word.

But I could imagine what a specialist meant. For example doctors who specialized in particular area, like ENT, Cardiology, Orthopedics etc. Patients sought the specialists when they felt that their case was beyond the generalists or based on the generalists advice.

It was much later in my corporate career that I came across generalist and grew an appreciation for generalists. My understanding was that a generalist is someone who has a wider perspective, has varied experience and manages the whole business with multiple functions and deliverables. I aspired to become one and ended up doing varied roles and multiple functions.

Definition

Generalist	**Specialist**
Less competent in more	More competent in less
Knows less of more	Knows more of less
Less skilled on more	More skilled on less
More width and less depth	More depth with less width

Beyond the definition we make many general associations with the generalist as well as the specialist.

Generalist

- Jack of all and master of none
- Character that can fit anywhere
- All-rounder
- Competent in several fields
- Can do anything
- Overall in-charge
- No defined role
- Big Picture people
- General Managers
- Problem Solver

Specialist

- Authority
- Subject Matter Expert
- Deep domain focus
- Functional expertise
- Little knowledge beyond their specialization
- Lack big picture
- Works in a silo
- Expensive

Value

This is where a world of generalization and specialization differentiates! We pay relatively more for specialty restaurants and specialty products. We ask people what they specialize in. Hence in general a specialist has a higher probability to have a higher perceived value. However this is not universally true and would be different at different situations.

For example a start-up company values generalists, simply because they cannot afford specialists. When start-ups hire generalists they can do more with less number of generalists as they double up in other areas as well unlike specialists.

However, an established and growing company would value and invest in more specialists. The generalists are less likely to be hassled by uncertainties and they are of higher value at uncertain times and situations. When you select a teacher you may value a specialist but when it comes to a mentor you may value a generalist.

Some differences

Generalists solve problem at hand - they look for solutions for the problems at hand and tackle problems that involve specialties beyond theirs.

Specialists give solutions at hand – they look for problems with the solution at hand and tackle problems bounded by their specialty

Generalists are the ones who can assume the role of a specialist that is needed for the situation.

Even a General Manager is a specialist as he or she is a specialist in managing the overall business involving multiple outcomes.

Reality

No one is a 100% specialist since in reality one would always be a partial generalist.

In a corporate scenario as you grow in your responsibility, the very responsibility may make you a generalist as you are called to do more. Also that most people will have secondary interests or would be forced to do beyond their specialization and hence no one is a 100% specialist.

Similarly no one could remain a 100% generalist as you are expected do many things well and at least one thing extraordinarily.

The perception of you will always be relative to start with. You will be perceived as a specialist if what is seen of you is higher than what they have in them. If you have relatively higher knowledge on something you are credited to be a specialist.

The equation tilts when others knowledge goes higher than yours or the environment are changed. Once specialist will always be a specialist is not true! Let us say you are specialized in selling to consumer. When the consumer buying behavior changes and if you have not kept pace with your learning you no longer will be a specialist.

What do businesses need?

Being a specialist is how one would be known, valued, hired, or promoted. Specialists do very well when the future is known and predictable.

But the business world is evolving so fast that the future is becoming more uncertain and ambiguous. Hence future plans will no longer work for decades or even years. Businesses want to plan for the future that is known and hence is shrinking the definition of future itself. Soon the measure of future would be in quarters or months since any future plan may not work beyond that period.

Hence businesses still need specialists but these specialists need to remain specialists by staying relevant. That is possible only when one keeps learning and become a specialist on more than one thing.

Specialists will continue to be in demand. If one is a specialist in marketing to consumer he or she is more likely to be hired by companies that have consumer-focused businesses. While one is a specialist in marketing to consumer and if you have also learnt enough about social media marketing, selling online one would soon have become a specialist in online sales and social media (where we would have started as a generalist).

Potentially we could now be hired by a Business-to-Business (B2B) company who wants to leverage social media and online sales as they may perceive us as a specialist.

It would be easy to become a specialist in the area of adjacencies and many times it may become mandatory to become a specialist in these adjacencies to remain a specialist in the core.

For example, when one is selling to consumer, social media and online sales are mandatory to know. The adjacencies that one is specialized in could become adjacencies or even core for some others and that opens up opportunities.

Simply put keep yourself updated, renewed and changed to stay specialist and add more specializations to stay relevant, be valued, be recognized and remembered.

It is worthwhile remembering the 80/20 rule. Be 80% specialist and 20% generalist to have a pipeline of future specialization. This will help you have a broad view and also make you multi-dimensional.

Summary

In the final analysis, I feel that it is not about choosing to be a generalist or a specialist. What is important is to do something in our area of interest and earn enough to support the lifestyle that we are comfortable with so that we enjoy what we do and that gives us the opportunity to stay specialized in the general area of interest.

When we are in our area of interest we will have the interest to learn more within that to stay relevant and remain a specialist for the known future. Is it better to have one great friend or many good friends? I would say it is best to have few great friends but to keep many good friends as well. In the same way specialization in more than one (with general knowledge on many) would not let you down. My best advice is to consider oneself a generalist and to learn continuously and become more of specialist over time.

About the author:
Joe Francis lives with his family in Bangalore. He is a freelance Consultant &
Independent Director. He worked for twenty seven years at multinational corporations
before striking out on his own. He was a hosteller at PSG College of Technology and
graduated with a degree in metallurgy.

Question & Answer

Is Employment better or Entrepreneurship?

As an Intrapreneur, I had opened up new business streams for my organization, built an India sales organization from scratch, made it immensely successful in India, ASEAN and Australia through direct sales and channel partner sales.

However, I am going to be self-employed. No more straddling on two boats. Time has come for me to take the boat to immense success.

Krishnan
Production Engineering
Bangalore

REFLECTIONS

The Ethical Dimension of Living

It is a truism that when we were graduating, technical content loomed large; because that is what defined us and got us started in our careers. However, as we progress in our careers, another dimension looms large; and that is the consideration of ethics. The author traces his early exposure to ethical living and explains how the moral dimension needs to be a supreme consideration in our lives.

Introduction

I am from a mining township in Neyveli. I studied at a Christian missionary convent run by the *Sisters of Cluny*. Here I learned the importance of *Moral Science*. Looking back, when this subject was being discussed, I was very young and did not appreciate all its dimensions. But as I have traversed my career, and look at society around me, these moral dimensions loom large and above many other priorities. The school helped collect money and supplies for disaster relief including when the Bangladesh refugee crisis took place. Thus, helping others and making a difference in their lives was an early learning and example for me.

Both my elder brothers were engineering graduates, so it was somewhat in a family tradition that I take the same route. Given my scores and ranking I merited a seat in the metallurgy department. Many fun experiences of life in the hostel included learning to live on one's own away from the family, learning how to behave in larger groups of peers and occasionally kicking up a fuss to get better services at the hostel facility.

Job on the line

Upon graduation, I had an offer from Carborundum Universal which I gladly accepted. However, a surprise was in store for me. For whatever reason I failed a course in my final semester! This had not happened to me during my entire tenure and I was faced with this incredible news.

Upon learning of this situation, the company gave me the option of continuing work and giving me a grace period of 6 months to take the exam again and pass it.

This was the first ethical dilemma I faced in my professional life. Do I continue the job which I was given on the assumption that I had cleared all my exams, now that the company learned of this fact?

I decided that this was not an ethical thing to do. I said to the management that *I would rather quit my job, pass the exam and re-apply for the job* and take the risk that the position might be filled by another candidate.

Both the company and my professor were taken aback that I would take such a stand. Later when I met my professor Dr. N.K. Srinivasan and discussed the situation and my decision he said that I made the first right ethical choice in my career.

It is cold in the north

While back in college attempting to pass my exam, I had time to spare and was thus able to investigate other job opportunities. Through a friend, I got a job interview in Delhi at his company. Upon arriving there, I found myself with neither warm clothes nor appropriate attire for a formal interview. My friend in Delhi helped me get ready.

I did well in the interview and was offered a job. But again to my surprise, the offer was even further north in the state of Himachal Pradesh.

I had to meekly admit that this was not at all my expectation and that I had to go back all the way to the south, discuss the offer with my family and then revert with my decision.

It all worked out well in the end. However, the lasting impression I have from this episode is that when *one is out of their element, it is very difficult to anticipate what the business and cultural requirements might be.* And in such a circumstances, it is difficult to keeps one's head straight and make rational decisions.

However, thanks to the help of my friend in Delhi, even though I was indeed out of my element, I was able to make the right choices both tactically and strategically.

The difficulty of being good

I would like to share an important book on this topic written by Gurcharan Das. The title of the book is, *'The Difficulty of Being Good.'* This book explores the question of ethics and moral dilemmas in a very compelling way. The author runs through the backbone of the Mahabharatha epic quickly, so the reader is up to speed on what the story is and who the main characters are. Even for a reader like me who perhaps is familiar with the epic, this was a good refresher. It is written in a 'lets get to the point quickly' style, but as a reader you realize that it is a palate cleanser for the complex meals to come ahead.

In the author's take on the epic, the central event appears to be the episode of Queen Draupadi's humiliation in the King's court. One would normally assume that this particular scene was a dramatic episode. It is assumed that the central scenes from the epic were Arjuna's dilemma or even Karna's demise. What the author is exploring is the question of *Dharma*. A central question of Dharma is posed by the humiliated Queen to the assembly of Nobles. What is the role of each Noble sitting in the court watching this humiliation? Why were they silent?

The author brings vitality to the scene. He asks such penetrating questions of the characters assembled. He builds so many layers to the answer. Not only do we need to know what the right thing to do is, but also need to understand that it is not easy to decipher it.

A question of ethics

One of the most famous courses online on ethics is by Prof. Michael Sandel at Harvard University. This course is called *Justice* and it is available on *YouTube*. I highly encourage watching this as it zeroes in on the questions of ethics and what is right.

As I listened to Prof. Sandel, a fundamental set of questions occurs to me.

For example:
• Is there anything such as 'morality' at all – or is morality just a thin layer of rationalization in a world that is really a jungle?
• Should a society help old and frail people? Should society help sick? Should society help orphans?
• Does a privileged person have more rights than a non-privileged person?

What is a just society?

We live in societies and have elaborate laws, rules and customs codified over thousands of years that govern our behavior. There are two ways of looking at this. One is, is there a way in which we can organize ourselves where the greatest good can be provided to the greatest number of people (not 100% good to 100% of people - that is impossible). That is the 'utilitarian' view. In this equation, you can ask in general are people in

Ethiopia better off than people in Somalia? Another way of asking the same question is, regardless of outcome, are certain things right, and other things wrong?

For example, do I have right to own property? In war time certain rights might be taken away, such as due process. But is right to property fundamental? In this regard is the capitalist system better than socialist?

We know there are universally held truths such as: Parents want to care for their children (even among animals). Or that people want to protect their homes and their families. People want to be on a winning team (read: growing economy). A good system will take these positive impulses and construct around it, while minimizing the damage from the darker side of human nature. It is a question of balance - how do you balance the greater good vs. individual rights?

Summary

Throughout my career as both a student and a professional, I have been struck by the question of ethics both large and small. From watching relief measures for Bangladeshi refugees as a child to making job related decisions as a person starting my career, I have seen questions of ethics at every turn. These questions, unlike technical questions, do not have formulas. One has to dig deep into one's cultural, moral and spiritual depths to find the right answer. But what makes life rewarding is wrestling with these questions both personally and for our society. As the great civil rights leaders in the United States Martin Luther King said, *"The time is always right to do what is right."* In the same vein, I wish that all of you look deeply at the question of ethics as you start and make progress in your professional careers. It is only then that we will live in a just society.

About the author:
Subramaniam graduated with a degree in Metallurgy from PSG College of Technology. He has worked in various parts of India and in the west. He currently lives with his family in Bangalore.

REFLECTIONS

Working in America

Many graduates today come to America either on assignment or for careers. What does it feel like to land on those shores? The author recounts how his story of finding his way through an enormous economy, and indeed finding a place in dynamic Wall Street.

Introduction

Like all of my classmates from the class of 1986, I emerged with an undergraduate degree from the PSG College of Technology, with a major in Production Engineering. I was a budding engineer, eager to take on the world with all the knowledge gained from the four precious years at one of the best colleges in India.

For this twenty one year old fresher, the term 'production' meant any or all of the processes that began as a requirement, brainstorming of ideas to cater to that requirement, transforming those ideas into a blueprint, developing a model or prototype, testing and finally rolling out the finished product, which was typically something that could be perceived by the five human senses.

True to that perception, when Ashok Leyland - then the second largest automobile manufacturer in India, offered me the position of an Executive Trainee, I lapped up that opportunity with great enthusiasm. Soon, we 'engineers' would be rotated around various departments as part of the Induction Training. We took pride in learning about thousands of components in a chassis and their manufacturing processes. We also used to quiz each other on these details, deriving petty pleasures.

From Axles to Pixels

As part of the Training, we also had a 3-day visit to systems department - the only 'clean' environment where everything was in immaculate condition, thanks to the needy computers that always required air-conditioners in operation. Back in those days, some companies used to refer to these departments as MIS (Management Information Systems) or EDP (Electronic Data Processing).

Every one of us was drawn towards working in that atmosphere, where the folks seemed pretty cool, cranking out reams and reams of COBOL code and talking in buzzwords - none of which made any sense!

There was an announcement for one vacant position and they had all of us (trainees) appear for an aptitude test for the selection process. With my measly knowledge of computers or programming languages, I had every reason not to take the test - but I did! The even more surprising part is that I got selected! *Moral: Be careful what you wish for.*

Thus began my accidental journey of Application Software Development, that started with a manufacturing company and has been navigating through various interesting stops along the way like banks, brokerage firms, media, entertainment and healthcare industries.

Which boat are you in?

The general trend of Indians coming over to the US as students underwent a significant change in the early 1990s with the government starting to shift focus on 'Specialty Occupation' Visas - in which people with development skills in the latest software were mostly sought after. Towards end of 1993, this demand escalated to such an extent that companies were training people rigorously with the intent of meeting the software development requirements of the US companies.

The events that unfolded during the mad rush were spectacular! As with any other phenomenon, it created a situation whereby screening became essential at every stage to weed out candidates with fake credentials and dubious claims.

The consular officers at Chennai were rejecting the H1B Work permit applicants at random - pushing everyone to the edge. I recall an experience undergone by one of the female applicants from the company I worked for during her Consulate Interview, a process that was never considered to be technical.

The officer looked at her application and asked her '*What is this - Sybase?*'. The girl said that it was a relational database and went about explaining it in detail. The officer wouldn't budge. He slipped a post-stick note across the counter and asked her to write a program right on the spot, asking her to explain the various functions! She had such a harrowing ordeal.

My interview, in comparison was a non-event and within a couple of days, I found myself boarding my flight to the US for the first time.

Glimpses into the early stages

New York's winter greeted me on the morning of March 6th, 1994 at John F. Kennedy (JFK) Airport, as the North East United States was recovering from the repeated assaults of seventeen snow storms.

Just as in movies, I was expecting someone with a placard to get me to my company guesthouse but there was no such thing. Nobody bothered or solicited me for a cab ride. Everybody was minding their own business - making no eye contact!

One of the Human Relations personnel from my company in India had given me a fax number, insisting that it was a phone number and that was all I had - not even the address to my destination! Being Sunday, the company address was useless. Here I was, with two suitcases, with all the energy and enthusiasm suddenly snuffed out of my body, feeling like a complete fool in a strange place. Welcome to New York!

After subjecting me to a decent lesson on patience, my ride came in a human form around night with a panic ridden face, profusely apologizing to me. En route the 75-mile ride to my place of stay in New Jersey, it felt surreal taking in the view of the Statue of Liberty and the spectacular Manhattan skyline. I was exhausted but greatly relieved, as I had to report to my client the next morning at 8 AM at the famed Wall Street!

Ain't over...yet. It has just begun!

The next morning, I reported to my client - Bankers Trust, in Manhattan, NY. With the frustration of the previous day and sleep deprivation pushed aside, my thoughts were swirling around the prospects of working in one of the skyscrapers alongside the wizards of the financial district.

The west's recognition of the talent pool from India has come a long way over the last twenty years. Folks that came from Asian countries had to get established themselves culturally and emotionally in order to get acceptance into the mainstream American system, and this process could take anywhere from a few months to years, depending upon the individual.

For the past several years, companies in India have been indoctrinating US-bound employees with everything they need to know regarding the work and culture right down to intricate details, including the slangs and accents. This has helped bridge the gap to a great extent.

My client manager - a petite woman, received me with a smile, and asked where I had been (I later learnt that my company had given her the notion that I was already living in the US for some time). When I replied 'India', I saw the look on her face change abruptly. I would've said the same even if my company had asked me to tell a lie. She immediately gave me an impromptu written test in 'C' programming language, hoping to see me fail so that she could get rid of me. *I passed the test but failed to erase her prejudice.*

A little commentary on prejudice. In the globalized workforce, many cultures and traditions are mixing. What corporations are recognizing is that we may all carry *'unconscious bias.'* That is those exhibiting the bias are not aware that they are exhibiting the bias. Thus, there are corporate programs being put into place for managers in particular to be trained to detect in themselves and in their colleagues, propensities for such behavior and take corrective action. This is indeed a welcome development.

In any case, I started my work at Banker's Trust, in spite of the prejudice exhibited by my hiring manager. I took a deep dive into getting to know the processes and applying my knowledge. Two weeks later, I was shown the door anyway - no reasons!

Clients enjoyed the try-before-hire privilege on people similar to products. The timing (and not the episode) bothered me a bit since my wife and 5-month old daughter had just joined me then. However, this initial failure only increased my resolve to be honest, although I realized that it may not act in your favor always! *Being honest may have initial setbacks but eventually it wins.*

Moving on...

Within a couple of weeks, I cleared up a round of interviews and landed as a consultant with Merrill Lynch, Somerset, NJ where I had my first taste of Corporate America.

I would like to mention some noteworthy facts - bound to make today's IT person chuckle: Windows 3.1 with 8 MB RAM was the prevailing operating system on the PC that I got at first, which was eventually upgraded to 16 megabyte Windows-95 was still being developed at Microsoft! One of my colleagues was bragging about his computer working on a Pentium processor with 640 megabyte hard disk space!

Other than the humble beginnings in technology, the work in itself was engaging and despite being a so-called *'fresh-off-the-boat'* immigrant, I had excellent relationship with every one of my colleagues who belonged to a culture quite different from my Indian roots.

They would drive me on numerous occasions to the train station (as I didn't have a car in the beginning), accompanied me on afternoon walks in the beautiful woods in the company campus, and took me out to occasional luncheons. These are all some of my most cherished memories from my early days at work.

New York Bound (Again)

After nearly a year, one of my contacts kept pursuing me to join his company to work for a media conglomerate - Viacom Inc., in Midtown Manhattan. With that began my tenure in New York City that would last over eighteen years in various spheres of applications development using a wide range of software technologies, serving Media/Entertainment companies and theme parks. I witnessed several mergers and acquisitions, floating of an independent Information Technology company and also company splits!

Similar to Lower Manhattan's popularity as the financial hub, Midtown has its own share of stately high-rise buildings with the added charm as the center of entertainment, always teeming with tourists. Broadway shows are popular year round. The Empire State Building with its observation deck atop its 102nd floor, offers amazing views of the city and the neighboring states. During fall and winter, the whole city comes aglow with a festive atmosphere, offering a welcome break from the depressing short days. Rockefeller Center has a tradition of Christmas tree lighting every year where in a crew is engaged in selecting the best Norway Spruce that meets their dimension specifications. Central Park is an oasis of greenery set amid the concrete jungle.

For nearly two decades, Paramount Plaza and Rockefeller Center were my prime places of work, the former, located near Times Square. But I was not crazy enough to visit the area on New Year's Eve!

A lot of interesting, life changing events happened during this time and I would like to list five among them. I relate to them as part and parcel of my working in this country - by being directly involved or as an observer.

1. Y2K Readiness and Preparation - It all boiled down to the early stage programmers (perhaps in the 1960s and 1970s) trying to save 'precious' disk space by opting to store year in two digits as opposed to four. Sure they might have had brainstorming sessions on "What happens when the year becomes 2000?" But I'm guessing that those who raised that question were 'shot' down with the message "*Let's deal with it when it comes!*"

And when it did, what a phenomenon it created in terms of a glut of resources pouring in from all over the world! Some people believe that the Y2K bug was a hype but I subscribe to the belief that we averted major issues due to our preparedness.

2. Dot com Explosion and Stock Market crash - How could anyone forget about the Information Technology bubble that burnt investors when it popped during March 2000? Established IT personnel were jumping ships at a moment's notice to join technology start-ups with grand 'promises' of stock options and several got burnt. The US Federal Reserve Chairman Alan Greenspan probably had a premonition when he gave that speech about three years before the bubble burst with his famous quote on "*Irrational exuberance.*"

If only they had heeded his words!

3. September 11, 2001 - Just a short distance from my office in midtown Manhattan, on that morning, the whole world was witnessing the best and worst acts of mankind and the fragility of human life all at once. The world changed forever!

I recall the words of our CEO in his calm voice: "*Today is a sad day for America. I request you to evacuate the building. I don't know what else to say but wherever you go, be safe.*"

But the city's resilience was evident when the trains started operating within less than four hours following the indescribable tragedy. I boarded the second train home. It was packed but there was deafening silence in the train, a mental state I would like to call as 'Controlled chaos'.

People were looking towards the direction of the smoke from where the twin towers stood only a couple of hours ago, wondering but not daring to ask "*That's my place. Where do I go to work tomorrow?*" - still coming to terms with why/how they survived.

Over the next decade, like a Phoenix from the ashes, from the place close to where the Twin Towers stood the Freedom Tower (One World Trade Center) would soar 1776 feet into the sky - the number symbolizing the year of US Declaration of Independence.

4. Pagers - Mobiles - Smart phones - What's next? - Picture a few people conversing with each other at work, when one person suddenly reaches out to her pocket as though an ant bit her. She takes a quick look at that thing she gets from her pocket and scrambles to a nearby phone to call that number. That seems so ancient! Yes, these conversation distracters have come a long way from the pagers of the early 90s to the Smart phones of today!

5. Security and Preventive Measures - With problems associated with viruses, hacking, and system crashes, major headway is being made in the areas of Security, Disaster Recovery and Information Protection.

What do Americans look for in the Information Technology work force?

Technical knowledge is a given for an IT professional. Although it helps to be a specialist in a certain area, it is more important to be adaptable to new set of software and methodologies, sometimes, to the extent of foregoing one's expertise to learn a whole new tool, as dictated by factors such as market demand and company policy changes. Here are three things I list which Americans look for at any time in a Technology specialist.

Bang for the buck

In general, Corporate America has become highly value-conscious in terms of spending money on their human resources. *A lot of emphasis is given to the concept of Return on Investment (ROI).*

Gone are the days of reckless spending and jet-set lifestyle of executives, thanks to the waves of recessions the country has undergone and the increasing scrutiny by investors. While every effort is made to select the right candidate or a contracting company for an Information Technology project without compromising on quality, the odds generally seem to work in favor of the lowest bidder. Significant time and money is invested in writing up the SOW (Statement of Work) with the participation of heavyweight lawyers to safeguard the company interests from Contractors with poor standards. Hourly contracts have been gradually taken over by fixed price projects.

Trustworthiness

During the transition period between Merrill Lynch and Viacom back in

January 1995, I had to work in both places for a few months. I used to get back home from my work in New York City and drive to my old client in New Jersey - about 50 miles apart. It was exhausting to do it every day and logging in remotely was not prevalent in those days.

My client manager at Merrill rented out a laptop and had me work from home at my own time. Not only that, he took the responsibility of filling out my consulting time sheet hours and submitting it to payroll solely based on my verbal input, just to make sure that my pay was not interrupted. He relied on my words and had no way of monitoring my work. I'm still grateful for the tremendous trust that he placed in me that made him go out of the way to help me.

Communication

Indians have a distinct advantage over other nationalities, thanks to their English proficiency, but that's not the only aspect I'm referring to. It's important to use discretion in communication. *Nothing can put off a business manager than a list of technical jargons that a programmer uses in his conversation.* Additionally, succinct exchanges are quite common in a fast paced environment. Therefore, the challenge for the IT professional is to know to speak the right language in the applicable forum.

The Globalized worker

As the world is getting smaller by the day, the working nature of a typical Information Technologist has undergone a major change. Virtual Private Networks have pervaded the work atmosphere, bridging not only the distance gap but also the time gap. Gone are the typical 9-5 work days.

Everyone is now equipped with the extended leash to their office so that they are able to stay in touch with their peers or managers whenever needed. As the layer between personal time and work time has started becoming more of a blur, it has become increasingly important for the IT worker to maintain a work-life balance. I see this pattern only continuing in the foreseeable future and therefore, I list the following traits that would effectively define a Globalized worker:

1. *Cross Trained* - to be able to support multiple areas rather than one.

2. *Dispensable* - so that another person could fulfill his role when he takes off on vacation or quits the job.

3. *Multitasking* - Seriously? Isn't multitasking just for computers? Sadly, this commonplace term has enveloped today's IT worker as they are expected to juggle between conference calls, emails and chats - all at once! While some people claim to be good at that, it compromises to a great extent on efficiency. *How about adopting Mindfulness instead?*

4. *Readiness to share knowledge* - People would agree with me that not everyone in this field is forthcoming, 'job security' being mentioned as one obvious excuse. The global worker has everything to gain by realizing that no one can operate in vacuum and by demonstrating the eagerness to help.

5. *Short learning curve* - be able to get up to speed quickly.

What happened to the Production Engineer?

In Information Technology terminology, production is a common usage to denote an active or live environment. Going back to a generic software development life cycle, a requirement is conceived, a system proposal is made, which in turn is accompanied by analysis, design, and construction, before it is subject to QA, culminating in implementation whereby the program is released to the live environment.

While comparing this with the life cycle in my Introduction section, strangely enough, I do see a parallel existing between them. The finished product in this case is not something perceived necessarily by the five senses, rather mostly by the mind playing a crucial role.

Summary

I have recounted my personal journey making my way to the United States and my career and lessons learned along the way. New York should be counted as one of the greatest cities in the world. It is a dynamic, multi-cultural melting pot. Indeed one can even say that the culture and dynamism of New York is distinct from the state of which it is a part and even the rest of the country. The pop music singer Billy Joel sang it best when he crooned '*The New York State of mind.*" A stint in New York will give the global professional a taste for capitalism unleashed and meritocracy valued.

Likewise, every city in the US has got its unique flavor that symbolizes its character, as defined by its history, local food, sports teams, points of interest and landmarks, giving its inhabitants their own bragging rights.

From the day the pilgrims landed centuries ago, this country has come a long way in terms of its culture, rights and religious freedom. Living and working in the land of the free, one gets to appreciate in the truest sense the three inalienable rights namely life, liberty and the pursuit of happiness - which together form the essence of why this country was born.

About the author
Suseendran (Hari) works in Software Stewardship and Asset Management division of a Health Care company in Philadelphia. He was a hosteler at PSG College of Technology, graduating in Production Engineering. He lives with his family in Columbus, New Jersey.

Question & Answer

Which leader do you admire the most?

Emperor Ashoka. Here is an example of a leader and uniter, who forged a nation and built its infrastructure. At the peak of his power, he learned the most important lesson about life and spirituality - on the battlefield. He is a real-life historical figure (not mythological). And after learning his abject lesson, spread the message of peace far and wide.

H.G. Wells wrote of Ashoka: "Amidst the tens of thousands of names of monarchs that crowd the columns of history, their majesties and graciousness's and serenities and royal highnesses and the like, the name of Asoka shines, and shines, almost alone, a star."

Surya Kolluri
Mechanical Engineering
Boston, USA

REFLECTIONS

Business Life in Dubai

As is evident, the authors in this book are currently based all over the world. Is it enough to have good professional skills to succeed everywhere? Or are there other important factors such as language, culture and customs? Dubai is a global melting-pot and a great place to explore this critical question.

Introduction

I was in my seventh of eight grade I was fascinated by computers, thanks to a popular writer, who went by the pen name *Sujata*. His story '*Sorga Theevu*' (Paradise Island) portrayed a small totalitarian island nation run by a huge machine where every inhabitant was so controlled they believed they were happy. In the story, when the machine develops a glitch, the manufacturer wouldn't want to fix it, and as the only available alternate expertise an Indian Engineer is summoned to perform the task. I wanted to be that Engineer!!

I was determined to study Electronics and Communication engineering as I thought if offered the closest to what held my fascination for computers. then. My father studied engineering at the College of Engineering in Guindy, and remembered that one of the interviewers was Dr. G.R. Damodaran, who happened to be the Managing Director at PSG College of Technology. Thus my father knew of and had great reverence for him and the institution he ran.

Even though I had admission to other college's including my father's own alma mater, I chose PSG College of Technology, since I got the admission there earlier, and was smitten …

My career begins

I was hired by the British Physical Labs (BPL) on campus. However, I immediately sensed issues with the work environment and my immediate manager. I decided not to continue that role within four months of joining. I returned back to Chennai and looked for a role in a hard-core telecom company. I started in a research and development capacity and then became a project manager managing the deployment of telephone systems in rural India. This gave me a chance to travel all over the country and had a wonderful time. Without doubt, this was the most exciting period of my career.

The United Arab Emirates and Dubai

The region Gulf, at the forefront of which is the country UAE (United Arab Emirates) and its fabulous city Dubai have all been in the midst of an unprecedented growth. Dubai is a very new city and well in to the 20th century was nothing more than a fishing hamlet with pearl diving tradition.

The city, which originally ventured out as the hub of Gulf, is now the hub of Middle East and can stake claim very soon as THE HUB.

Sounds preposterous?! Not if you have been witness to the transformation for some twenty years like I have. It is a place that seems to be at odds in its elements and the same elements go towards synthesizing this place in to what it has become - a true melting pot.

On to Dubai

Going to Dubai, was just fated I think. Working or living in Gulf was never an idea for me. It is a place where people went to earn some quick money at the cost of professional progress, is what I used to think.

But a telephone call one evening in February 1993 changed all that. My senior in the Indian company who had moved to Dubai a few months prior called over phone and persuaded me to join him. I said yes even though I wasn't sure.

Once I was in Dubai, I didn't give myself more than two years, given my pre-conceived notions about the place. Now, I write this note, twenty two years later!

My classmates have moved to many parts of the world, including New Zealand, Asia Pacific, the Middle East, Europe, Africa and the Americas. My aim in this chapter is to describe life in Dubai, so that, in addition to get a flavor for experiences in professional life, that have been described throughout this book, the reader can also get a flavor for geography and culture as well.

Arriving in Dubai

It was 1993. After flying nearly four hours from Chennai (called Madras then), the landing announcement was made. I had a window seat and when I looked down all I saw was a vast expanse of golden sand. I was prepared for this, but still the view of the desert was quite extraordinary. In the breadth of the golden desert the odd dark streaks –the roads- stood out. Strikingly, they were very straight lines. Of course in a plain desert one could build a road without being bothered about negotiating a hill or a water body I thought.

Where is my job going to be in this desert?

Soon enough I started to notice buildings and more roads. Then I saw the blue sea as the aircraft bent and turned. Not a big town I thought. I was left with a sense of desolation. Was it really a wise decision to come to this place for work?

My senior colleague who was instrumental in getting me in here had told me that I would be picked up at the airport. I was checked in to a hotel and after a shower the phone rang. It was AKS, my senior colleague in India and my boss in the new company too. When I listened to his usual, cheery *'Haaaiii Venkat'* I was enlivened.

Many of us got together that evening, and as a toast was proposed toast, I remarked I would have never thought alcohol was allowed in Dubai. His answer was one of the many strange answers that I have come to learn over the years. He said, non-muslims can drink, so long as they are well behaved!

Starting in Dubai

One of the first things I learned about work in Dubai were the strange timings of work. When I walked into work on my first day, I was told that work starts at 8 am, which was fine by me.

But then I learned that people go back to their homes for lunch at 1 pm. They would get back to work at 4.30 pm!

The offices then close at 8 pm six days a week. However, the last day of the working week is Thursday. And on Thursday's we worked from 8.30 am to 2 pm. However, these rules were for the private sector. The government worked on a different schedule. Their work hours were from 7 am to 2 pm six days a week.

After having spent several years in India, traveling everywhere, and used to a different work schedule, this aspect of work-timing was what struck me immediately. *The pace of work, the hours of work and the rhythm of work can be different in different parts of the world.*

The other difference that struck me immediately was the heat in Dubai.

It was HOT when we stepped out of the building for lunch. Actually, HOT is not the word. I asked what the temperature is like and I was told it could be 49 degree centigrade. I landed in the country on 4th of July. Temperatures in July and August can soar beyond 50 degrees centigrade, and as per local laws the outdoor work must halt if the temperature crosses 50.

Oil field and construction, which is predominantly outdoor in nature of work, are the major employers of labor force. In the afternoon when it is extremely hot, they had a break from work and this was even true of those people who had desk jobs in the days when the air-conditioning was not yet common.

Thus my two observations are intertwined. The work hours were driven by the situation with the weather and heat. The historic tradition was being maintained even in nineties were every work place and every home is air-conditioned.

However, with globalization, and after twenty years of being there, now the work timings have all changed to be regular now. No more afternoon naps. Most organizations have switched over to 5-day week and we now have Friday and Saturday as weekends.

Oh ... These names!

Another aspect I noticed as interesting was names of people. Sue Haddad was the name of my CEO's assistant. It seemed like almost every second Lebanese person had the surname Haddad. (now I know I was wrong then). There was then a Syrian who was referred to as Arbaab. I learned that Arbaab means owner, boss and is an equivalent of expression *malik* in India .

Every office I walked in to, every patio that was there to be straddled on had the portraits of two people in the minimum and sometime up to four. They were the pictures of the president and the prime minister. H.H. Sheikh Zayed Bin Sultan Al Nahyan and H.H. Shaikh Maktoum Bin Rashid Al Maktoum.

I asked my Palestinian colleague who was with me, how long Mr. Rashid had been the P.M. of the country?

He said the P.M. was not Shaikh Rashid, but Shaikh Maktoum.

I was confused. Maybe I would use the second name, like saying Mr. Clinton, for example.

He gave me a patronizing look and said if you say Shaikh Rashid, then it refers to his father and not to the PM.

I understood. It is Shaikh Maktoum son of Rashid of the Maktoum clan.

That is how the nomenclature works.

Women did not change their maiden name after marriage (not only in the Middle East, but in general in Islamic societies, women keep their maiden name).

In the case of women it was *Bint* instead of *Bin*.

This was all tremendous learning for me. *How people are addressed, how they wish to be addressed and what subtle differences might entail in a new culture and environment, were all important to understand.*

Clothing and attire

The attire of men is called *DishDasha* or also *Kandurah* a long white gown that extended up to the ankle but should not conceal the ankles. While almost 99% of the men wore white robes, other plain hues were allowed as well. The men covered their head with a scarf mostly white that was held in its place by a cord which sat like a ring on top.

Women wore black head to toe dress. They were supposed to cover their face as well with the cloth that covered their mane. They were not supposed to exhibit any ornaments or jewelry.

I was told that women wore very colorful bright silk clothing inside the black cover and within the privacy of their homes and with their men they did not have to use the black robe to as a cover. They also wore expensive jewels. This was evident when you saw the young UAE children who were not required to cover themselves. They wore bright, embroidered long gowns and flashy jewelry. UAE women while they covered their heads, most did not cover the face.

When you stepped out in the streets you not only saw white attired men and black robed women but more of western attire. It is because the country's population was made of large proportion of expatriates and the country also thrives as a place for tourists. Some 80% of the population were expatriates.

Conducting Business

This expatriate population most of who had a resident status that was renewed every three years had a mixture of people from the Sub-continent, the Far East Asians, Arabs, Iranians, Europeans (mostly British) and Americans.

Most large organizations had a British boss in the early days. Now Dubai's corporate leadership landscape has diversity of nationalities as does its overall workforce which has every nationality of the world represented.

In the first few months, I had to contend with wry British humor when negotiating a sales contract. On the plus side it hits you on the face and you learn being objective. And you also learn to respond with some sarcasm. And then you learn that the better people within the same group were excellent guys. And then you learn it is true of any group.

When you dealt with locals on the other hand -who were the business owners- it was another thing. They were very cultured and the business was extension of the culture. They would never ask you any difficult question within the proposal that you are putting on the table. They would never tell you what is incomplete in the proposal. In India we were used to being told in the face that the proposal in under engineered. You then rectify and you could win the contract. Here the local raised no objection and still you ended up on the losing side.

And when you WON it gets tougher even. When you delivered the project, you would discover to your horror the expectation was something else. You cannot contend that it was not in the scope. You were supposed to know. You are given business on trust.

This is true of any place. *but in this part of the world TRUST above all is key.*

Dubai Tomorrow

The population of the country has grown four fold in the last twenty years, but the composition is not much changed in terms proportions. Today Chinese form a good percentage and they were practically absent 20 years ago. Indians have assumed top positions in many organizations especially banks and technology companies.

There are many Indian owned businesses, which are top companies in the region. Amongst them are a couple of large multi-billion dollar enterprises started from scratch by Indian expats from very humble backgrounds. All of which goes to evidence UAE and in particular Dubai as the land of great opportunity and hospitality.

It has been astonishing how much Dubai has grown and changed in the last twenty plus years I have been here. It has grown from a small provincial place to a truly global city. It has become the destination for many of the global conventions. The work and culture has become globalized. Indeed, for our 30th anniversary, our class picked Dubai as the central place to meet.

When I think about Dubai and its aspiration, it feels like they would like to be the *Center of the World*.

I highly encourage young professionals to come and visit Dubai and see for themselves how they are transforming themselves, the region and indeed having an impact on the world.

About the author:
Venkat Raghavan studied Electronics and Communication engineering at PSG College of Technology. He works in the Telecommunication and IT services sector and lives in Dubai.

Question & Answer

Who inspires you?

At work, a manager of mine has been a strong inspiration. He was the symbol of hard work, clear thinking and cool temperament. He reached almost the top of the organization, no mean feat, and retired recently.

I feel I am fortunate to have had the chance to observe and study such a personality, and make changes into my attitudes and life, so one day, I can get there, and ensure that one top post remains efficient and does what it is meant to do.

As for what inspires, I think of Bharathiyar, "*Ucchi meedhu vaan idindu veezhugindra podhilum, achamillai.*" Also, the prayer: "*Gnana vairakya sidhyartham biksha dehi cha parvathi; Mata cha Parvati Devi, Pita Devo Maheswarah; Baandava Shiva Bhaktyascha, Svadeso Bhuvanatrayam.*"

Swarna Ramesh
Mechanical Engineering
Chennai

IV. WAY FORWARD

Trajectory of an 'Unreasonable' Man

In this section on the Way Forward, the author of this chapter takes inspiration from the trajectory of a great leader who was passionate about inspiring young minds. The author of this tribute narrates how this leader is an inspiration to all.

Prologue

The sky was a riot of myriad hues of red and gold. There was a cool breeze, carrying the whiff of the ocean. The sleepy town on the east coast of southern India was just coming alive. The boy had no time to enjoy nature's drama, he was waiting, eyes and ears keenly open for signs of the oncoming train. He had to collect the newspaper, run around the town and deliver.

The year was 1940. Daily life continued fairly as usual in spite of the World war that had been declared. But it brought enormous pressure to the family of this little boy of age eight. His father found it extremely difficult to manage the family with five children, his brother's family and the elders since they all lived as a joint family. Being a boat-man, who ferried pilgrims his entire income depended on the pilgrims to the temple town. Due to the war, people were not travelling much. With the rationing system, everything was scarce- food, clothes and all essentials. His grandmother and mother had to squeeze every resource to keep the family running.

The children were given food first and the elders were partially starving to manage the situation. The boy could hear his parents talking, after the lights were off, about the acute food shortage and need for more money. Strangely enough, the war gave him an opportunity to support the family soon. As an austerity measure, the government had removed the train stopping at the town. The newspapers used to arrive by the early morning train and it became a challenge. His cousin, who was running the newspaper business found a way out. His affiliate would come in the train and throw the bundles on to the platform as the train chugged on. But some one was needed to be there at the platform, collect the newspapers and then distribute. The boy accepted to do the job happily – he was overjoyed that he could make a difference.

Delivering newspapers

From his biography we learn that this was really tough for an eight year-old boy, with all other commitments that he had already. He had to be at his Maths teacher's home for tuition at 4 am in the morning. He should have taken bath by the time, as per his master's wish – which means getting up even earlier. After the Math class, his father would take him to religious lessons and morning prayers. Then he had to run to the station to collect newspapers and then around the town delivering. He knew people by the newspapers that they read. Soon his customers started a few friendly words with the charming boy. He had to reply with a smile and keep running. He had to be at the school by 8 AM, after the sparse morning meal that his mother made sure he had every day. It did not end there. In the evening he had to go around collecting money and settling the accounts. He had to oversee the boatmen and repair works on the boats. After that he would sit by the kerosene lamp and start his studies for the next day.

The whole experience made him learn a few vital things. He learned to estimate time to run with the newspaper bundle, to each locality, so that the papers were delivered reliably at the same time every day. He learned to mentally remember the money due from each customer and maintain accounts. These were simple but profound lessons for a great project manager as he would turn out to be. He learned that to keep a commitment, even as simple as delivering a newspaper, one had to be up and ready to face the day, whatever else may happen. Homework, tuition, prayers, all carried on, but the train would not wait. He had to be at the station at the right time and at the right place to catch the bundles as they came flying in. He never complained about the ordeal he had to undergo. *He loved every moment of it, as he was inspired by the bigger cause of supporting his family.* That became his nature. The unreasonably passionate boy learnt his life lessons and became a great leader.

When the country paused

India witnessed a singular phenomenon recently. TV Channels forgot the usual political topics and switched the prime hour debates. There were posters in every small town and village, with condolence messages. The fourth estate allocated front page real estate that is normally reserved for more interesting profiles than a scientist. The entire country was mourning, paying homage to Dr. A.P.J. Abdul Kalam. His was a life worth lived, worth emulating. Dr. Kalam , son of a boat Man, the newspaper boy who studied under kerosene lamp, turned in to a rocket scientist, led the Indian missile program and finally became the President of India. There lies the story that captured our imagination.

Indian Space Research Organization - ISRO

I was at ISRO myself, early in my career and had a great time at ISRO. I worked on the project to send first Microprocessor to Space from India. It was a thrilling moment when the Software that I had written was working in Space and sent data back to earth. I had the great fortune of working under eminent leaders like Prof. U.R. Rao and Dr. Kasturi Rangan.

Dr. Kalam was one of the few scientists who started with ISRO in its early days. Today ISRO stands as a symbol of what India could achieve – with its recent Moon and Mars Missions. India joined the elite space club having sent a spacecraft to our neighbours in space. India became the first country to launch a successful Mars mission on maiden launch. The magnitude of the achievement is unparalleled – it was done at a fraction of cost compared to others.

The entire mission was planned to be at a specific time window to have minimum energy spent. If the November 2013 window was missed, the wait would be another two years. It was launched as planned within the window. The spacecraft travelled 780,000,000 kilometres to reach Mars. It took about 10 months. In between precise navigation and many manoeuvres had to be done. The communication system was an achievement by itself. The signals take several minutes to reach the ground station – which was specially constructed with a 32 metre Antenna. But such frugal innovation and self-reliance became the DNA of ISRO from the start, thanks to people like Dr. Kalam.

Interestingly, ISRO had a humble beginning in 1962. The early attempts of Rocket launches started in a tiny fishing village called 'Thumba,' on the west coast of India, almost near the tip of the Peninsula. Thumba was chosen for the sounding Rockets program as it was nearest to the magnetic equator. The facilities were appalling by today's standards. The scientists did not even have a building. They had to use a Church and the priest's home to store the rockets and conduct meetings. Rocket parts were carried on hand and on bicycles.

ISRO into the future

Why does space have such a hold on us? There are many answers such as how it advances science and engineering. But it is best to share a quote. When asked by The New York Times why he wanted to climb Everest, British mountaineer George Mallory, who died on the mountain during his third expedition there in 1924, famously answered, *"Because it's there."*

As we know from recent news, ISRO is going great guns. It has been a trailblazer. ISRO has achieved unparalleled success with Moon and Mars missions, soon there will be manned missions and reusable space vehicles.

Space has always captured human imagination and the exploration will continue.

A Dream Fulfilled

As the project manager of SLV, India's ambitious Satellite launch vehicle program, Dr. Kalam asked for a team of 150 people and got a team of 50 people. But he managed in his usual spirit. India had many other priorities than Space Research, but he could convince the Prime minister and continue pursuing his dream.

President Kalam can be aptly described as an unreasonable man. He was a man of *unreasonable* dreams, he was able to inspire his team to share his vision. He was not grounded by harsh realities. He could have become a boat fleet owner following his father. He could have become a newspaper agent.

But what made him struggle to study? He could have joined a Government department in a clerical job after graduation, but he continued to study Aero Space Engineering. He took up Space Research. He took up the challenge of making India's first Rockets to launch Satellites.

When I think of President Kalam, I am reminded of a quote by the playwright George Bernard Shaw: *"The reasonable man adapts himself to the world; the unreasonable one persists in trying to adapt the world to himself. Therefore, all progress depends on the unreasonable man."*

Summary

In this anthology, each classmate of mine has shared his or her trajectory – and these stories are from all over the world and are indeed full of useful lessons and insights. Throughout this anthology, other classmates have answered the question on who has inspired them. In this chapter, I have combined those two approaches and traced the trajectory of legendary figures who inspires us all. I hope as you construct your respective career trajectories, you take inspiration from this story as well.

About the author:
SriKrishnan graduated with a degree in Electronics and Communications Engineering from PSG College of Technology. He is Vice President Engineering, leading the Business Unit in Car Multimedia, Vehicle safety systems and Engineering Tools in a MNC. He started his career at ISRO, working on Satellite Telemetry and Data Handling Systems. He lives with his family in Bangalore.

Question & Answer

Which leader do you admire the most?

Gautama Buddha. He was a great man of conviction, who had the courage to question the validity of the caste system in India and to challenge the Hindu religion. He is the reason Hinduism mellowed and changed many of its tenets.

As a prince, he had all the luxury in life, but he left them behind. Taking up renunciation and leading a movement, what he accomplished is incomprehensible.

Truly, he was the man with Godly qualities.

Shoba Dharmalingam
Electronics and Communication Engineering,
Chicago, IL, USA

A 2016 Commencement Address

Universities in the west, in particular, have pageant filled graduation ceremonies. At these ceremonies, a guest speaker delivers a commencement address. The author here presents such an address in this section on the 'Way Forward.'

Introduction

To be candid, I had no idea what a commencement speech was when I graduated from PSG College of Technology three decades ago.

However, thanks to the Internet, I have read or watched and have been inspired by commencement speeches by many famous personalities. A commencement speech is usually made by a notable figure during the graduation ceremony. The person giving this speech is the commencement speaker. A famous example is the commencement speech by then Apple Inc. CEO *Steve Jobs* at Stanford University in 2005.

At some point of time though, I started wondering whether all these "feel-good" speeches were really impacting the bulk of the graduating class or was it aimed at the elite few at the top of the class who are perhaps eventually destined to make a big impact. How would it be to have a commencement speech made by somebody who was perhaps not-so-famous? What would, for example, I say, if I ever got invited to make a commencement speech?

I decided to use this anthology being published my class of 1986 as an opportunity to indulge in my unfulfilled desire (*niraiveraatha aasai*).

So here goes…

Your journey begins

At the outset, hearty congratulations on successfully making it through all those up's and down's that are an inevitable process of any academic journey. You have now perhaps reached what may seem like a destination at the end of a journey. But in reality, you are now at what certainly is a starting point of a new journey.

Let me try and over the next few moments share my perspective on what such a journey will look like and what are some of the forces that act on you when you make this journey. I am going to ask you to use your imagination as I describe your upcoming journey.

I would like each of you to think of yourself as a little ball just about to embark on a journey. There are two dimensions in this journey - one dimension will be left or right and the other dimension will be up or down.

There are many forces acting on this ball (i.e. you). Let us start with the inevitable that each and everyone of us face without exception. That is a pull to the right, that none of us can really do anything about. The ball will roll at a steady pace to the right, as often as the earth goes around the sun - that is, the march of *"time"*.

Let me introduce one more object now – a little bar or line that defines what you would like to achieve. Let me call this the *"success line"* or *"dream Line"*. This position of this line is for you to set and for sure this is something that can be re-set over time. You can choose to set this bar very high or adopt a step-by-step approach. The good news is that this is your line or *"your dream"* and it is for you to decide where this line should be. In other words, your definition of success is what matters.

You really should not worry too much about what others think about the position of the line and how you would like to "alter" this position over time.

What you certainly need to do is pay attention to the other forces that are likely to aid your upward and downward movements.

Forces to contend with

Let us start with three terms that we constantly keep hearing:
- Drive
- Passion
- Talent

How do we understand these in simple terms? These are certainly three significant forces often confused for and arguably similar.

I look at these forces as
- Internal or
- External

Passion
This is an irresistible external pull that attracts us in a particular direction. This could be the force that inexplicable urge to hop onto a gym first thing

135

every morning – come rain or shine. This could also be the urge to drop whatever you are doing and reaching for your camera when you come across something that appeals to you.

Drive

Drive on the other hand is that internal push that "drives" you towards achieving what you seek to achieve. In other words, it is that upward force that takes you closer to your "*success line*" even as you keep moving towards your right dragged by the force of time. This is that endless urge to complete the Sudoku puzzle or crack that code, come what may.

Talent

Talent is something that some of us are fortunate to be born with or in some cased gifted with. Maybe this is genetic – or maybe something that was dormant and discovered through some external force. There is no denying that this is a huge asset if this is something we discover in the course of our journey else we could end up as one who never really leveraged his or her talent.

Impact of these three forces

Let us look at the impact of these three forces on our journey. For sure, your upward move towards your "*success line*" will be vastly accelerated if you are able to marry your passion with your talent and your drive to be successful.

Unfortunately, not all of us enjoy this happy marriage! An example of someone who perhaps falls in this category, to use a sports example from the game of Cricket, is *Sachin Tendulkar*. There is no denying that he is a talented cricketer. And he was certainly passionate about the game and wanting to stay in the middle as long as he could. That did not mean he gave up his practice sessions to keep building on his talent. And the results show what he has achieved.

The progressive achievements during his career reflect the constant re-setting of his "*success line*" to achieve something higher with the passage of time.

Not all of us are talented though – or at least may not discover our talents right away.

What if we were not talented, or we think we are not talented, or are told we are not talented?

I would argue that we should not lose hope. There are many out there who may not have the same degree of talent but have achieved success through their sheer drive.

Drive that manifests in the form of perseverance and hard work. A sportsperson, to continue using Cricket as a metaphor, who I can think of, is *Rahul Dravid*. For sure he may not have the same degree of talent that *Sachin Tendulkar* demonstrated. However, that did not stop him from achieving great heights.

At this point, I cannot help recall a very well-known line of one of the greatest innovators of the twentieth century, Thomas Alva Edison who said, "Genius is 1% inspiration and 99% perspiration."

I have always worried about the fact that I was unsure about what I was really passionate about. Or coming to think of it, was I passionate about anything at all! What if there was nothing that really pulled me in a particular direction?

Over time, I have stopped worrying about this and convinced myself, that one day, I would eventually discover my passion. And even if I did not, I always had the force of "Drive" that could propel me upward.

The most successful amongst us tend to understand clearly what our passion is and align our drive and leverage our talent, if it does exist. Some of us may however, have our passion completely independent of what we may be "driving" for – and this is not a bad thing. It only goes to show that we are multi-faceted in our approach and nurture what perhaps is a hobby.

One way of looking at this is almost like splitting our line of success into a *"Professional"* and *"Personal"* lines and pursing two different paths. Reaching either line of success gives us happiness – so nothing should hold us back from pursing our passion and at the same time driving towards success in the professional sphere – even if these are totally independent of each other.

And for those who may not yet have discovered what their real passion is – despair not – one day you will. Till then keep pressing with your drive to propel you upward.

The presence of and of the three forces of Drive, Passion & Talent tend to give you that upward force. However the absence of any of these does not necessarily mean that they will pull you down. It may lead you to "drift" over time – and unless you deliberately choose to do so, will not drag you downwards.

Luck and Health

That leads to a couple of other forces that we cannot ignore: Luck & Health.

Luck

These are two forces that can move the ball (i.e. you) be either upward or downward in life's journey.

It is undeniable that for some people are where they are because they were at the right place at the right time. This sort of luck is assigned in popular culture to former American President Bill Clinton. His victory of a sitting President George Bush (thanks to a third party candidate Ross Perot), his escaping serious allegations during his campaign, shenanigans while in office (thanks to radical opposition) allowed him to exit office as one of the most popular Presidents the country as ever seen. Had he not have such luck, he might well have been driven out of office; indeed, he is the only the second American President to have been impeached by the U.S. Congress.

This can be attributed to a chance occurrence or simply because they had that inner voice telling them where they should go. It is great if we can "get lucky" but it is foolish to rest on getting lucky as a strategy to get to our success line.

Health

For obvious reasons, it is important that we pay attention to Health – since it is something that we have control over. I have bundled Luck and Health in the same argument with a reason. Take care of your "Health" and "Luck" will take care of itself.

Indeed, it is gratifying to see that health consciousness is all pervasive today in terms of what we eat, focusing on the right amount of sleep and taking care to do exercise. Thanks to wearable technologies, we have immediate feedback loops that help us monitor our state of well being (at least physically speaking).

Let me take care to add, by health, I mean both physical health and mental health. Indeed, inspite of Rene Descarte splitting the mind and body in the thinking of Western Culture, I would wholeheartedly ask you to hew to the

Indian mindset of *yoga* - which is the holistic fusion of mind and body.

The external forces

Let us now talk about two external forces. Your "detractors" and your "influencers".

Detractors: Detractors are those that are always and almost certainly with an intention, pushing you down. Recognizing these detractors and avoiding them is certainly an important skill one needs to develop – because no matter who you are and what you are up to, there will always be detractors.

On the other hand there are going to be your well-wishers who care for you and will provide you with the necessary impetus that you need to move ahead. While this advice may sometimes seem unsolicited, we need to recognize that these are people who have the right intentions of helping you and not harming you. The least we can do to reciprocate their intentions is to hear them out. You of course, have the option to decide whether you would like to follow their advice or otherwise.

Influencers: It would make sense to pick up from amongst these influencers, individuals who can coach and mentor you. A coach is one who can provide you the advice to overcome specific shortcomings. A mentor on the other hand is one who understands you and takes a holistic approach towards your personal and professional development.

Sri Adi Sankara sang, *"Sat Sangatvae nissangatvam, Nissangatva Nirmohatvam,"* meaning through the association with good people comes detachment to false ideals. Legendary business leaders or sportsmen – whether it be General Electric's Jack Welch (incidentally GE was founded by the aforementioned Thomas Edison) or India's own Sachin Tendulkar, even in the prime of their career, depended on coaches and mentors to iron out their imperfections and make them more effective.

On the topic of coaches it was fun to see two coaches in the recent final of the U.S. Open tennis tournament. In the men's finals, tennis legend *Roger Federer* was playing another legend *Novak Djokovic.* However, what caught my eye were their respective coaches. Federer's coach was *Stephan Edberg.* And Djokovic's coach was *Boris Becker.* When I was your age, and just graduating out of college, those two coaches were in the finals of several majors, and they are legends as well!

These influencers can be seen as those little booster motors that give you that extra power needed to jump.

The safety net

Let me now introduce an additional object in this picture. A "*safety net*". This net is your Family, Friends and Professional Network. They form an integral part of your personal and professional life. You need them in good and bad times. In bad times, when the forces drag you down, they literally serve as a safety net to prevent you for crashing to the ground and help you bounce back.

Network: This could be through emotional or financial support or through any other method that can help rescue you. In good times, these are your well wishes who can give that extra bounce to leap forward with speed. Growing this network over time helps you make that net even wider. At the same time, not nurturing your network and keeping it alive will result in weak threads giving way in the event you land on that net.

Please allow me to share with you a rule of networking that if you learn and practice will serve you well over your careers. In every project you are assigned, wherever in the world you are and whatever task you are assigned - try and make 3 friends in that assignment. By friends, I do not mean, you worked well with these people during the project. I mean, make a *connection*.

This connection should transcend the current project work you are doing. The connection could be about geography, sports, hobby, arts - anything that is genuine and is of mutual interest. Let us say you are given 1 or two assignments a year. Over a period of 10 years, you will have made 50 - 60 professional friends. And remember, they are also progressing in their careers, just like you are. Lo and behold! Suddenly you are in a senior position after a couple of decades, and you have *connections* that you can lean on (even after the specific project is long forgotten).

Knowledge and Learning: In your journey, you also need a lens to look at that unchartered territory that you are heading towards. This lens is nothing but Knowledge. This is not just an in-depth knowledge of your field but an understanding of the environment in which you are operating. We are in an era where the pace of change and new technology will present new challenges to overcome.

As one of my friends has written in this anthology, education is very different from learning, and the latter is the most important. Thus, learning and knowledge can help us convert these challenges to opportunities. This additional object in the picture emphasizes on the need for you to constantly keep learning in your journey.

Summary

Today, you are at *Ground Zero* of what you should see as an exciting journey into unchartered territory. It is up to you to set your target. These targets could relate to your professional or personal sphere. And of course, these targets are moving targets that you could change over time. It is up to you to control them – or else somebody else will be controlling them.

There are multiple forces that could aid you in your journey. These include Passion, Drive, Talent, Luck, Health, Family, Networks, Influencers, Mentors & Detractors.

Navigating these unchartered regions requires you to keep constantly learning. If you develop an understanding of the forces and use the upward forces to your advantage and keep learning continuously, nothing can hold you back from *success* or from *living your dream*.

From the class of 1986, my and our best wishes as you embark on this journey.

About the author:
Mr. JC Sekar has just moved out of the corporate world to "leap across the chasm" and pursue entrepreneurial ambitions related and leading to a Safer, Smarter, Healthier & Sustainable world. He lives with his family in Singapore. At PSG he was a hosteller and graduated with a degree in metallurgical engineering.

Question & Answer

Which Leader do you admire most?

Bharat Ratna JRD Tata. For succeeding in business while maintaining high ethical standards, for managing the huge conglomerate of diverse enterprises, for the excellent philanthropic charities across the globe, and for being a genuine, compassionate leader.

Bharani Rangabashyam
Electronics and Communication Engineering
Boston, USA

Our World in the Future

In the 1980s, one of the seminal books in popular culture was 'Future Shock,' by Alvin Toffler. Thirty years or more later, the book holds up well and was rather prescient. The author of this chapter, who is an expert on Public Policy, makes a similar attempt to observe the current cultural, technological and economic trends and describe what they might mean for our future.

Introduction

As kids, most of us were familiar with the *'Dippy Bird.'* Between the capillary action at the beak and the weight of the fluid between the neck and the bottom of the dippy bird, the mechanism would cause the bird to oscillate and dip its beak in the water once in a while. This example helps make the point that when enough variables come together, it can make a system move dramatically. As I scan the world and note the amazing progress in various fields, it occurs to me that we might be approaching our own dippy bird moment in human affairs. In this concluding chapter, I will try to trace these trends and describe what they might mean for us.

Manufacturing and Supply Chain

The Economist magazine had a cover story on 3-D printing in a recent issue. They have been beating the drum on 3-D printing for nearly a decade now. I finally got curious and decided to do a little exploration, being an old manufacturing hand myself. I came upon a *YouTube* video demonstration of 3-D printing that was quite extraordinary. A real tool, a wrench, was actually replicated. It is about four minutes long and really worth watching.

The apogee of modern manufacturing is the perfected, gigantic global supply-chains epitomized by *Walmart + China*. The US based retail chain, *Walmart* is a gigantic outlet selling the products at the cheapest possible price by squeezing suppliers and making the supply chain efficient. China, as we have all come to know, has been dominating the manufacturing side of things. But it is now popularly known that the value, for example, in the case of the Apple *iPad* is in the majority, in the design and the applications. The manufacturing cost of the *iPad* itself is in the approximately 10% range of the product.

Imagine a world, where each of us has some version of this 3-D printer in our garage. We own a car worth tens of thousands of dollars. So it is possible to imagine a replicator at that cost. We would be paying for design of a product and replicating it ourselves (a sweater, as an example). In the least, one can imagine, local 'printing villages' where everything is produced.

In this world, the current global supply chains would be radically altered. Rather than shipping entirely assembled products, perhaps more raw materials would be shipped. Today's manufacturing and global supply chains have perfected what Henry Ford started. Tomorrow, we might finally realize a series of interconnected villages, in a way fulfilling Gandhi's vision - but in a 21st century way.

Education

To this day, we are all striving for and aspire for our children the best education at the best institutions. A degree from a prominent school, such as the Ivy League in the US is much sought after. That is seen as a stepping-stone. But right under our noses, information, education and know-how is exploding through multi-media - be it podcasts, YouTube lectures or TED Talks. In writing about Salman Khan, the creator of the Khan Academy, Bill Gates, the founder of Microsoft, has written one of the most prescient articles about education. Another example is Harvard Professor Michael Sandel's lecture on Justice. This video that is available on YouTube, has been viewed more than four million times! It is one of the most powerful lectures on ethics and moral reasoning one will ever hear.

In such a world, what is the value of college education - particularly when the cost of education has been outstripping inflation multiple times every year for nearly two and a half decades? At some point, the notion of an expensive college education will be challenged by society, since a lot of content is all around us anyway.

An important consideration here of course is the quality of the content. A highly ranked academic institution, is so ranked for a good reason – similarly a degree or certificate. The ubiquitously available content will only gather value and validity upon being properly vetted, validated and recognized.

Environment

Not a day passes, when my son does not remind me of some aspect of my behavior that is not environmentally friendly. Between mining, dumping and garbage creation, we have been flooding the planet with waste - including nuclear waste. At the same time around the world we have been denuding forests at an alarming rate. Glaciers are disappearing in front of our very eyes.

But environmental consciousness has been rising so that if not in our generation, in the next, we could well see the return to more preservation and less exploitation. With accelerating economies in China, India and Brazil there won't be enough oil and coal left to fuel rapid growth. There is no question that out of sheer necessity we will need to shift to solar, wind and other renewable resources.

Most corporations are already getting serious about '*green*' and the carbon footprint. Imagine a time, when one is in a meeting, and no one brings any paper. Think of a scenario where the U.S. President or the Indian Prime Minister issues an edict just to his executive branch: no paper in any meeting. Imagine the U.S. Government or the Indian Government stopping the use of paper. This is not inconceivable that this could happen.

Of course, a moral dictum (such as 'print less paper') pales in comparison to the impact technological change can have. The digital revolution and its impact through ERP systems and digital signatures provide a significant opportunity, for example, to cut down paper use.

Loving work and working on what we love

Work as we know it today was epitomized by Gregory Peck in the movie *'The Man in the Gray Flannelled Suit.'* In the past a company man was the order of the day: loyal to his organization, making a work-life balance choice. With the world economy in the condition it is, there is increasing necessity to find work that best suits one's skills. The imperative is building a career that best fits your needs. For example, our son's ski instructor, during the summer season, is a racecar driver and during fall, an automotive consultant. That is NOT a man in a gray flannelled suit!

For two decades now we have been talking about the transition from a manufacturing to a service economy. We are now headed from a service to a knowledge economy. With service being delegated to robots that can do everything from keeping your house and refrigerator in the right condition to vacuuming your house. This is not far-fetched at all. Just think about the number of tasks you can do now just with our iPhone or Android that required service workers: banking, shopping, reading and listening. Well *Siri* the automated voice on the Apple Phone is not there yet. But any doubts that *Siri X.0* is round the corner with a lot more capabilities?

Longevity

Human longevity has increased by leaps and bounds and it is no longer surprising when we regularly cross 100 years in age. That is 35 years past traditional retirement. In fact traditional retirement is near being retired. The notion that we might live after the age of 65 worked until that point gives rise to reasons to re-think everything from financial planning, to activities and responsibilities.

With regard to financial planning, it seems to me to be pretty clear that it is not enough to have a 'number,' i.e., assets in the portfolio when one reaches a pre-determined age. What is important is infusing it with meaning from the past and the future. Experts suggest three key questions to ask:

Question 1: What was it like with money growing up for you?
Question 2: If you had a calamitous emergency today that wiped out your savings, what would you do?
Question 3: If you had all the money you need today, what would you do?

With the population in the oldest segments growing fastest around the world, policy makers are going to be pressed to come up with ways of addressing new issues - such as delivering elder related health care services on a large scale.

Genetics

We are unraveling the 'toolkits' within the gene (and not just human genes) that are responsible for various things including growing organs. Imagine organ farms that grow organs that we need. Sean Carroll is a Professor of Molecular Biology ('EvoDevo') at the University of Wisconsin. He has an excellent podcast on the topic of synthetic life.

We are creating micro-organisms from scratch that can follow instructions and do various tasks for us. Craig Venter, the pioneering US geneticist, heralds the dawn of a new era in which new life is made to benefit humanity. This would be a range of life forms starting from bacteria that churn out biofuels, soak up carbon dioxide from the atmosphere, to those that even manufacture vaccines. He created a new organism that is based on an existing bacterium that causes mastitis in goats, but at its core is an entirely synthetic genome that was constructed from chemicals in the laboratory. The single-celled organism has four "watermarks" written into its DNA to identify it as synthetic and helps trace its descendants back to their creator should they go astray.

As reported in the Guardian newspaper in the United Kingdom, Julian Savulescu, professor of practical ethics at Oxford University, said: "Venter is creaking open the most profound door in humanity's history, potentially peeking into its destiny. He is not merely copying life artificially ... or modifying it radically by genetic engineering. He is going towards the role of a god: creating artificial life that could never have existed naturally."

Craig Venter feels that this is an important step both scientifically and philosophically and that this progress has certainly changed my definitions of life and how it works.

A vision for the future

It is of course tough to predict what date and time these will converge and what shape the converged outcome might take. But one can sketch out a vision, *a strawman.* Imagine in the future we will choose to live in smaller communities. We won't go to college, and we won't work 9-to-5. There won't be large factories spitting out products. There won't be defined careers either. There won't be any mega stores to go to. Neither will there be paper nor plastic much around. There won't be paper currency either.

Here, we would be living in a highly inter-connected world, plugging into knowledge and education uniquely tailored to each individual and paced accordingly. Value will be based on unique and new ideas contributed to human understanding and endeavor.

A designer would come up with an idea and it can show up at everyone's home electronically (perhaps even through holograms). We would pay for the idea and have it locally produced - maybe in your garage or in the village central - and most of the value would be for the design and not the material or the production. The information about the earth, its waters, its bio-diversity, the map of the universe with its galaxies would all be digitized and be at your fingertips (literally). You could communicate with anyone anywhere about what is happening (five years ago, my brother was able to send me a mobile message from the top of Mt. Kilimanjaro).

In this world, we would be free in the truest sense of the word, to pursue what we each individually like and derive value from the overall network of knowledge contributors (not workers) around the globe while enjoying a silent, virtual exchange of digital currency. Lest this be confused for a 60s commune, this is not 'give up everything' idealism. Rather it is a convergence of amazing knowledge, scientific progress and technology that will allow us to jump headlong into human affairs, but not be caught as a cog in a giant wheel.

In the future, 'Gone Fishing,' might not necessarily mean absconding from our responsibilities. It might mean we are doing the most natural thing in the world. While out there, by the brook, fishing rod out, with a gentle breeze swishing by and one's dog, Rover lolling in the sun, one might have an Emily Dickinson-ian insight (The poem *I'm nobody! Who are you?*, comes to mind). And oh, by the way, one could post it on the future generations' version of Facebook for friends, family and the general public to react. They might even credit one's bank account should they feel it has value!

Summary

Nanotechnology. Computing power. Digitization of information. Private space flights. Mining Asteroids. There are even more such trends developing around the world. Just like the dippy bird analogy that I used at the beginning of this chapter, these trends, both individual and societal are converging. However, let us acknowledge an ever-present caution. One does hope that the Yang, that will accompany this Ying, with bad guys having access to the same technology and knowledge can be kept at bay!

But looking, reading and absorbing these trends, it is pretty clear that we are nearing an apogee in human affairs. An apogee in the sense that we have pushed twentieth century organization, scale manufacturing, global supply chains to their limit and that perhaps now events will start reversing in an evolved sense and headed back to a perigee where things are more localized and customized.

In this *Brave New World*, traditional stages in life such as going to school, getting a job, staying at the job(s) until retirement, and then fading away into retirement may no longer be the norm. Individuals will need to take control of their education, re-skilling, networking, financial planning, health management and much more into their own hands. Each individual will need to navigate this ever more rapidly changing world with the new tools of the 21st Century. This is going to be fascinating to watch unfold.

About the author:
Surya Kolluri is a Managing Director at a Global Bank focused on Public Policy and Planning. He lives in Boston with his family. He graduated with a degree in Mechanical Engineering from the PSG College of Technology, where he lived for fifteen years. He has a masters in Mechanical Engineering from Drexel University and an MBA from the Wharton Business School at the University of Pennsylvania.

Question & Answer

Who do you admire the most?

If you ask me, I would say that I admire my friends who are leaders.

I have learned a lot from my friends and picked up qualities from them through all these years. It is better to admire people who are among you, rather than people whom you have neither met nor moved with.

Many of my friends are successful; and it is nice to know of these success stories and enjoy them!

M.L. Ashokan
Production Engineering
Coimbatore, TN

ACKNOWLEDGEMENTS

At beginning of the introduction to this book, we mentioned three classmates who started a discussion group exactly ten years ago. Clearly, we would not have written this book had they not started that group. Our thanks to Ashokan, Mohan Alapat and Gauthaman.

Our thanks to our two friends, Ganesh and Venkat Raghavan, who in 2014 spurred the creation of a book we created for the internal use of the class

One person conceived the idea for this book, formulated the steps, reached out to volunteers, motivated people and followed through diligently right up to the stage of completion. Our classmate, Surya Kolluri deserves special acknowledgement for his tireless efforts and for the extent to which he has been a positive influence on all our lives.

Our grateful thanks to each of the authors of this book, who have taken the time out of their busy schedule to contribute a chapter and or observations. Indeed, there is not book without them. The list of names is too long to name each author here. However, each chapter bears the author's name and ends with a brief biography.

Our thanks to Suresh Babu, Krishnan, S.R. Uma and Srikrishnan for offering suggestions and editing the book during the course of its development.

Our appreciation for Krish's (Krishnan's) wife Rekha and daughter Aishu for designing the graphics for the book cover and Surya Kolluri's son Rahul Kolluri, for proof reading the manuscript. Also, thanks to Bharani Rangabashyam for suggesting the title of the book and Suresh Babu's daughter Anisha Babu for selecting it from over fifty suggestions.

Our affectionate and joyful thanks to Prof. Radhakrishnan Nair, for writing the foreword. He was our professor while we were in college, later Principal of PSG College of Technology and Vice Chancellor of Vellore University.

Finally, our sincere thanks to several distinguished well-wishers who have taken the time to read this manuscript, and have given us testimonials to support this effort. Their names are listed with each testimonial.

Tracing the trajectory

www.ingramcontent.com/pod-product-compliance
Lightning Source LLC
Chambersburg PA
CBHW072304200526
45168CB00014B/383